I0489418

GUIDEBOOK of FINANCIAL TOOLS:
Paying for Environmental Systems

U.S. Environmental Protection Agency
Office of the Chief Financial Officer
Environmental Finance Program

August 2008

Guidebook of Financial Tools
2008 Revision

Table of Contents

Foreword ....Page i

Comprehensive Financial Tools

Section 1. Tools for Raising Revenue... Page 1-1

    1A.   Taxes ....Page 1A-1

    1B.   Fees and Special Charges...Page 1B-1

Section 2. Tools for Acquiring Capital .... Page 2-1

    2A.   Bonds ....Page 2A-1

    2B.   Loans..Page 2B-1

    2C.   Grants..Page 2C-1

Section 3. Tools for Enhancing Credit and Lowering Costs .... Page 3-1

Section 4. Tools for Building Public-Private Partnerships .... Page 4-1

    4A.   Public-Private Partnership Arrangements................................. Page 4A-1

    4B.   Public-Private Partnership Case Studies......................................Page 4B-1

Section 5. Tools for Delivering Financial Outreach....................................Page 5-1

Guidebook of Financial Tools
2008 Revision

Table of Contents

Targeted Financial Tools

    Section 6. Tools for Accessing State and Local Financing .... Page 6-1

    Section 7. Tools for Financing and Encouraging
             Pollution Prevention and Recycling... Page 7-1

    Section 8. Tools for Financing Community-Based Environmental Protection ...... Page 8-1

    Section 9. Tools for Financing Brownfields Redevelopment................................ Page 9-1

    Section 10. Tools for Financing Small Businesses and
             the Environmental Goods and Services Industry.............................. Page 10-1

        10A.   Equity Capital ....Page 10A-1

        10B.   Debt...Page 10B-1

Appendices

    A. Environmental Financial Advisory Board ....Page A-1

    B. Environmental Finance Center Network..Page B-1

    C. Environmental Financing Information Network..Page C-1

This 2008 revision of the Guidebook of Financial Tools is a reference work examining a wide range of different tools for financing sustainable environmental systems. The term "sustainable environmental systems" refers to virtually any successful or potentially successful environmental protection initiative, including public and private environmental protection programs. Environmental protection initiatives that were not previously sustainable can be made productive and sustainable with the proper financing. The ten sections of the *Guidebook* present outline information on over three hundred financial tools that can help make environmental protection initiatives more sustainable. This intensive revision includes the addition of a new section titled "Tools for Accessing State and Local Financing" which includes many state grant programs. The *Guidebook* is designed to assist all interested parties in the public and private sectors with finding the means of financing environmental protection initiatives that are appropriate for them.

Sections one through five of the *Guidebook* examine comprehensive financial tools, such as environmental finance organizations and websites, public-private partnerships, and traditional means of raising revenue, borrowing capital, and enhancing credit. *Guidebook* sections six through ten examine specialized financial tools, many of which are geared towards specific geographic areas and types of projects. These specialized financial tools include approaches to paying for pollution prevention, community-based environmental protection, and brownfields redevelopment. They also include ways of improving access to capital for small businesses and the environmental goods and services industry. Each financial tool in the *Guidebook* is divided into a "Description" section and a "Reference for Further Information" section that includes internet links and other references.

The *Guidebook* is the product of a collaborative effort among the United States Environmental Protection Agency's (EPA) Environmental Finance Program, which includes the EPA Environmental Finance Staff, members of the Environmental Financial Advisory Board (EFAB), the directors and staff of the university-based Environmental Finance Center Network (EFCN), and other contributors. The question of how to pay and who pays for environmental mandates is a central theme for the work of the Environmental Finance Program. The 2008 *Guidebook* revision and any future *Guidebook* revisions will remain as final drafts. One of the reasons for the ongoing "final draft" status is that the *Guidebook's* contributors are continuously discovering more unique and innovative financial tools to add to the publication. In this spirit, new financial tools will be periodically added to the online version of the *Guidebook*. The *Guidebook* is available online at www.epa.gov/efinpage and hard copies are available on request.

This section describes specific financial mechanisms which states and localities use to raise funds for environmental protection programs and initiatives. The following means for generating revenue are presented: general taxes, selective sales taxes, and fees. A tax is defined as a financial charge or other levy imposed on an individual or legal entity by a state or a functional equivalent of a state, such as a tribe. General taxes are levied on a very broad section of the general public, such as wage earners or property owners. Selective sales taxes are levied on the sale of particular commodities and services. A fee is defined as the price one pays as remuneration for services, such as government administrative services and utility services. Fees are also defined as financial charges for activities undertaken, including polluting activities such as solid waste disposal.

Many of these tools for raising revenue are used primarily by state and local governments, and some are used by the federal government as well. Revenues from taxes typically go into the general funds for state and local governments. The process of gaining voter approval for dedication or earmarking of taxes for environmental protection initiatives is often difficult, considering that government-funded programs vigorously compete for monies and the popularity of environmental issues rises and falls over time. Revenues from fees are often deposited into special funds related to the product or service upon which the fees are levied, such as fees on fertilizer and pesticide sales being deposited into a dedicated fund for pesticide and fertilizer regulation.

Taxes are by far the largest source of revenue for state and local governments. Fees, with the exception of user fees which raise significant revenue, are much less universally used and generate far less revenue than taxes. Some revenue generation tools are more suitably dedicated to specific environmental work than others. For example, large and relatively stable revenue sources may be ideal for environmental infrastructure capital and land-related projects such as parks, while smaller revenue sources can fund program operating functions such as personnel, monitoring, and technical assistance. Some taxes and fees have dual purposes in that they raise revenue in addition to acting as market devices to alter polluting behavior by requiring the polluter to pay for engaging in that behavior.

1.   Alcoholic Beverage Taxes
2.   Federal Fuel Taxes on Motorboats and Small Engines
3.   Energy Taxes
4.   Ecological Taxation
5.   Occupancy and Public Accommodation Taxes
6.   Insurance Premium Taxes
7.   Litter Taxes
8.   State Tax Check-off Programs
9.   Earmarked Selective Sales Taxes
10. Motor Fuel Taxes
11. Motor Vehicle Sales and Registration Taxes
12. Aviation Taxes
13. Real Estate Transfer Taxes
14. Rental Car Taxes
15. Tobacco Taxes
16. Open Space Sales Taxes
17. Taxpayer Contributions Toward Toxic Waste Site Cleanup
18. Gross Receipts Taxes
19. Corporate Income Taxes
20. Estate and Inheritance Taxes
21. Individual Income Tax Deductions
22. Local Sales Taxes
23. Tangible Property Taxes
24. Real (Ad Valorem) Property Taxes
25. State Sales and Use Taxes
26. Solid Waste Collection Taxes
27. Severance Taxes

Alcoholic Beverage Taxes

Description: Alcoholic beverage taxes are levied by states and the federal government on over-the-counter purchase of alcoholic beverages. They are based on volume or value of beverages sold, and include liquor, wine and beer. Since alcohol is distilled from agricultural products, the federal government or state governments could potentially justify dedicating a surcharge on alcoholic beverage taxes to agricultural runoff control or other land-based programs. Alternatively, since breweries require a large volume of clean water and discharge wastewater from distilling processes, revenues from an alcoholic beverage tax surcharge could be dedicated to drinking water treatment and point source water pollution control programs. Since administrative records of alcohol sales already exist, a tax surcharge would be administratively simple to collect and track. The demand for alcohol is relatively unresponsive to price changes; thus a tax increase would not necessarily cause a decrease in sales sufficient to offset revenues.

Reference for Further Information: The Tax Foundation Website:
http://www.taxfoundation.org/, search the Website on "alcoholic beverage taxes."
Minnesota House of Representatives House Research Website:
http://www.house.leg.state.mn.us/hrd/issinfo/ssalbvtx.htm. Tennessee's "Alcoholic Beverage and Beer Tax Guide," Nov. 2005: http://www.state.tn.us/revenue/taxguides/alcbevguide.pdf.

Federal Fuel Taxes on Motorboats and Small Engines

Description: Revenues from federal motor fuel excise taxes levied on motorboats and small engines are dedicated to boating safety and fisheries conservation related activities under the authority of the 2005 amendments of the federal Wallop-Breaux Act. The 2005 reauthorization of Wallop-Breaux redirects to the Sport Fish Restoration and Boating Trust Fund approximately $110 million per year of revenues from federal fuel taxes paid by anglers and boaters which previously went into the general treasury. The fuel taxes directed to the Sport Fish Restoration and Boating Trust Fund are distributed according to a formula supported by the American Sportfishing Association and a coalition of 33 other fishing and boating organizations. The Trust Fund amounts to about $570 million per year, which is directed to state fish and wildlife agencies as a primary source of their funding. Activities financed with the Trust Fund include fisheries monitoring, habitat conservation, and restoration; fishing and boating access facilities such as docks, piers, and boat ramps; and education and safety programs for anglers and boaters.

Reference for Further Information: American Sportfishing Association Website:
http://www.asafishing.org/asa/government/wallop_breaux.html. "Wallop/Breaux renewed & expanded," *Boat/US Magazine*, September 2005, available at
http://www.findarticles.com/p/articles/mi_m0BQK/is_5_10/ai_n15393958.

Energy Taxes

Description: Energy taxes are surcharges on bills for utilities such as electricity, heating oil, and gas. Many U.S. states and municipalities require utilities within their borders to charge energy taxes to their customers. When energy taxes are levied only on energy sources that emit carbon dioxide into the atmosphere, they are called carbon taxes. Carbon taxes make polluting energy sources such as coal and oil more expensive than renewable, non-polluting energy sources such as wind and solar. This creates economic incentives for consumers, power companies, and utilities to switch to renewable energy sources. Carbon taxes are not yet charged at any location in the U.S. The State of Vermont is considering implementation of a carbon tax. The debate over whether or not to adopt a carbon tax is currently most active in the European Union. New Zealand seriously considered adopting a carbon tax in 2005.

Reference for Further Information: Global Policy Forum Website:
http://www.globalpolicy.org/socecon/glotax/carbon/index.htm.
Myer, Rod, "Carbon tax too costly, says NZ," *Business*, December 30, 2005, available at
http://www.theage.com.au/news/business/carbon-tax-too-costly-says-nz/2005/12/29/1135732693442.html. U.S. Environmental Protection Agency Website:
http://yosemite.epa.gov/gw/StatePolicyActions.nsf/uniqueKeyLookup/BMOE5PVJHS?OpenDocument

Ecological Taxation

Description: Ecological taxation is a fiscal policy that introduces taxes designed to promote environmentally sustainable activities via economic incentives. This fiscal policy is referred to as the "green tax shift" in cases where it is designed to prevent changes in overall tax revenue by proportionately reducing taxes that are not believed to promote environmentally sustainable activities. "Ecotax" is short for ecological taxation. The taxes that are introduced pursuant to the policy of ecological taxation are called Pigovian taxes. Pigovian (also spelled Pigouvian) taxes are levied to correct the negative externalities created by market activities. Examples of negative externalities created by market activities include pollution from burning fossil fuels and increases in incidence of asthma in a particular geographic area as cigarette smoking increases in that area. Pigovian taxes levied for environmental protection purposes include taxes on products and activities that contribute to environmental pollution, such as taxes on motor fuels. These taxes encourage the consumer to use less of the product because they add to the price of the product.

References of Further Information: U.S. Environmental Protection Agency Website:
http://www.epa.gov/owow/nps/MMGI/funding.html#6.

Occupancy and Public Accommodation Taxes

Description: Occupancy and public accommodation taxes, frequently called "transient occupancy taxes," are levied on accommodations in hotels, inns, tourist houses, motels, and other lodging used by tourists, such as campsites and campgrounds. State and local governments use the proceeds from public accommodation and occupancy taxes for various purposes including environmental protection initiatives that can be beneficial to tourists. For example, Delaware dedicates 1% of its 8% public accommodation tax to its Beach Preservation Program, which is administered by the state's Department of Natural Resources and Environmental Control. Dare County, North Carolina uses 1% of the proceeds from its 5% occupancy tax to finance beach nourishment projects, such as the planting of vegetation and the building of structures including sand fences and dunes. These beach nourishment projects are done for the purpose of widening beaches to mitigate erosion from storms and prevent damage to property near the shoreline.

Reference for Further Information:
Delaware Code: http://www.delcode.state.de.us/title30/c061/sc01/index.htm.
Dare County, North Carolina Website:
http://www.darenc.com/depts/Taxes/collections/occp.htm. .

Insurance Premium Taxes

Description: Insurance premium taxes are levied on the state level. The revenues from insurance premium taxes are frequently dedicated to pension funds. There are a couple of ways that insurance premium tax revenues could be used for environmental protection purposes. First, insurance premium tax revenues could be placed in pension funds screened by managers to ensure that all capital is invested in companies, financial institutions, monetary funds, and/or other financial entities that have taken clear steps to minimize their environmental impact. Also, insurance premium tax revenues could be dedicated to specific environmental protection initiatives. For example, proceeds from taxes on auto insurance premiums could be used to fund air pollution control. Taxes on insurance premiums have a large tax base so they yield a significant and predictable revenue stream. Revenues from taxes on the premiums of mandatory types of insurance, such as auto liability insurance, are the most predictable.

Reference for Further Information: See "Green Investments" in Section 7 of this Guidebook. Federation of Tax Administrators Website, with links to the tax agencies of all 50 U.S. States and Puerto Rico: http://www.taxadmin.org/fta/link/.

Litter Taxes

Description: Litter taxes are imposed on businesses that produce, distribute, or sell consumer products that contribute to litter problems. These taxes are levied by state governments and municipalities. The State of Virginia levies a litter tax on manufacturers, wholesalers, distributors, and retailers of consumer products. Ninety-five percent of Virginia's litter tax revenues are used for litter prevention and recycling grants. The other five percent of Virginia's litter tax revenues goes to the Virginia Department of Environmental Quality to administer the grant program and provide support for the Litter Control and Recycling Fund Advisory Board. The State of Washington imposes a litter tax on industries that sell, manufacture, or distribute consumer products and packaging. Twenty percent of the proceeds from Washington's litter tax funds the Community Litter Cleanup Program, thirty percent of the proceeds funds waste reduction and recycling efforts, and the remaining fifty percent funds other litter cleanup efforts. The city of Oakland, California imposes a litter tax on fast food restaurants, retailers, and other businesses and uses the revenues from the tax to pay crews to pick up litter.

Reference for Further Information: The Tax Foundation Website: http://www.taxfoundation.org/blog/show/1362.html. Virginia Department of Environmental Quality Website: http://www.deq.state.va.us/recycle/programs.html. Washington State Department of Ecology Website: http://www.ecy.wa.gov/programs/swfa/litter/laws.html.

State Tax Check-off Programs

Description: State tax check-off programs allow taxpayers to "check-off" contributions to state programs on state personal income tax returns. All state tax check-offs, with the exception of those directed to political campaign funds, are donations from a taxpayer's refund. Many U.S. states use check-off programs to raise money for environmental protection initiatives. For example, Ohio's income tax check-off programs for endangered wildlife have helped foster the return of bald eagles and have helped preserve the habitats of rare plants such as the lakeside daisy. Oregon's tax check-off for wildlife funds habitat restoration and species management for the 88% of Oregon's wildlife that is not hunted, trapped, or angled. Kentucky's state income tax check-off program provides funds for the Non-game Division of the Kentucky Department of Fish and Wildlife Resources and the state's Nature Preserves Commission.

Reference for Further Information: Federation of Tax Administrators Website: http://www.taxadmin.org/FTA/rate/Checkoff03.html. Ohio Department of Natural Resources Website: http://www.ohiodnr.com/features/jan06taxcheckoff.htm.
Oregon Department of Fish and Wildlife Website: http://www.dfw.state.or.us/wildlife/diversity/tax_checkoff/.
Kentucky State Nature Preserves Commission Website: http://www.naturepreserves.ky.gov/helping/taxcheckoff.htm.

Earmarked Selective Sales Taxes

Description: Selective sales taxes, unlike general sales taxes, are applied to specific commodities. Any product or service could be subject to a state or local selective sales tax, provided that voter approval is gained in jurisdictions where it is required. Most selective sales taxes are either limited to a specific time period, or used to raise a specific dollar amount, and then ended. Any state or locality may seek to establish a selective sales tax and earmark it for a widely supported environmental purpose, such as parks, recreation, open space, nature centers and trails, and environmental education. The Coconino Parks and Open Space Program in Coconino County, Arizona is funded with a one-eighth of one-cent sales tax, which has been extended to 2012 and is devoted to the protection of natural areas and the creation and enhancement of county parks. Colorado Springs, Colorado's Trails, Open Space and Parks Program (TOPS) has used revenues from a one tenth of one percent sales tax, which has been extended to 2025, to preserve more than 3,100 acres of open space throughout the city.

Reference for Further Information: The Trust for Public Land Website: http://www.tpl.org/tier3_cdl.cfm?content_item_id=19601&folder_id=1365 & http://www.tpl.org/tier3_cdl.cfm?content_item_id=19602&folder_id=1365.

Motor Fuel Taxes

Description: Motor fuel taxes are imposed on the state and federal levels and are levied on gasoline and other fuels. All 50 U.S. states and the District of Columbia charge gasoline taxes. State gasoline tax rates generally range from 10 cents to 33 cents per gallon. The U.S. federal motor fuel excise tax (per gallon) is 18.4 cents on gasoline, 13.6 cents on Liquefied Petroleum Gas (LPG), 18.4 cents on gasohol, 19.4 cents on aviation gas, and 4.4 cents on jet fuel. Each time the federal motor fuel tax is charged, 0.1 cent per gallon goes to finance the federal Leaking Underground Storage Tank (LUST) Trust Fund. State and federal motor fuel tax revenues are typically dedicated to highway construction and maintenance. Revenues from state and federal motor fuel taxes could potentially be earmarked to fund air pollution control and related research.

Reference for Further Information: Internal Revenue Service Website: http://www.irs.gov/. Contact the Revenue Departments and/or Departments of Transportation for individual states. State of South Dakota Department of Transportation Website (has link to information on state gasoline taxes throughout the U.S.): http://www.sddot.com/geninfo_fuel.asp.
U.S. Environmental Protection Agency Website: http://www.epa.gov/OUST/ltffacts.htm. See "U.S. Environmental Protection Agency: Leaking Underground Storage Tank Trust Fund Grants" in Guidebook Section 2c.

Motor Vehicle Sales and Registration Taxes

Description: All 50 U.S. states charge taxes on the purchase and registration of new and used motor vehicles. Motor vehicle sales taxes are levied on vehicle sales and title transfers. Motor vehicle registration taxes are levied at the time of initial registration and registration renewal for vehicles. Registration is provided to document payment of motor vehicle registration taxes. Generally, the funds raised with taxes on motor vehicles are used to pay for highway-related state programs. Many states have statutory or constitutional limits on the earmarking of the revenues from motor vehicle sales and registration taxes. In states without these limits, a portion of the revenues from these taxes could be earmarked to air pollution control programs, public transit programs, bike trail construction and maintenance, or other environmental protection related initiatives. For example, the Illinois Department of Natural Resources Bike Path Grant program is funded through the state's vehicle title transfer tax, which is levied on cars, trucks, and trailers.

Reference for Further Information: Federation of Tax Administrators Website, with links to the tax agencies of all 50 U.S. States and Puerto Rico: http://www.taxadmin.org/fta/link/.
Illinois Trail Riders Website: http://www.illinoistrailriders.com/Horse%20Trails.htm.

Aviation Taxes

Description: Airlines and their customers pay a number of taxes and fees to a variety of authorities, both in the United States and abroad. The stated purposes of aviation taxes and fees mandated by the U.S. government include environmental protection, homeland (national) security, disease control, and airport operations and maintenance. Revenues from the U.S. federal Commercial and Noncommercial Aviation Jet Fuel Taxes and the U.S. federal Noncommercial Aviation Gasoline Taxes could potentially be used to fund programs to address air pollution caused by the burning of fuel by airplanes. Also, these aviation fuel taxes create an incentive for airlines to conserve fuel because they add to the cost of the fuel. The U.S. Federal Leaking Underground Storage Tank (LUST) Fuel Tax is an aviation tax that is used to raise funds for the U.S. Environmental Protection Agency's Leaking Underground Storage Tank (LUST) Trust Fund. The LUST Trust Fund provides states with grants for the purpose of addressing releases from Leaking Underground Storage Tanks containing petroleum.

Reference for Further Information: Air Transport Association Website:
http://www.airlines.org/economics/taxes/. See "U.S. Environmental Protection Agency: Leaking Underground Storage Tank Trust Fund Grants" in Section 2c of this Guidebook.
Global Policy Forum Website: http://www.globalpolicy.org/socecon/glotax/aviation/index.htm

Real Estate Transfer Taxes

Description: Real estate transfer taxes are charged to the buyer and/or seller of real property at the time of sale, based on a percentage of sale value of the property, a flat deed registration tax, or a combination. Sales of residential, commercial, and industrial properties are subject to real estate transfer taxes. Rates and dispositions for these taxes vary from state to state. Localities must seek state legislative approval in some states, such as North Carolina, before they can impose the tax. Some states make collection of these taxes a county responsibility while directing the revenues to the state general fund; other states give local governments the authority to collect and keep the tax revenues. Many states and communities use proceeds from real estate transfer taxes to establish dedicated funds for natural resource protection initiatives including the establishment of parks and preservation of open space. Transfer taxes can inflate real estate values and slow the market. However, dedication of real estate transfer tax revenues to popular land protection programs enhances the acceptability of the taxes.

Reference for Further Information: U.S. Environmental Protection Agency Website: http://www.epa.gov/owow/nps/MMGI/funding.html#6. The Trust for Public Land Website: http://www.tpl.org/tier3_cdl.cfm?content_item_id=1060&folder_id=825

Rental Car Taxes

Description: State and local sales taxes are frequently charged on car rentals; and the dollar amounts of these taxes are growing in many locations. State and municipal governments are increasingly using rental car taxes to finance civic projects. Rental car taxes could be used to finance infrastructure improvements, such as increases to the capacity of wastewater and drinking water treatment plants, that are needed to meet the needs of seasonal tourists. In addition, revenues from rental car taxes could be used to fund air pollution control programs. Rental car tax revenues might also be used to finance public transportation programs and projects. Washington State uses revenues from its rental car tax to fund its Regional Transit Authority, which is devoted to financing a high capacity, rapid public transit system.

Reference for Further Information: Yu, Roger, "Heavy taxes crash down on car rental industry," USA Today, available at: http://www.usatoday.com/money/biztravel/2006-04-18-car-rentals-taxes-usat_x.htm. The Tax Foundation Website: http://www.taxfoundation.org/blog/show/1172.html. Smarter Travel Website: http://www.smartertravel.com/car-rental/Beware-hidden-rental-fees.html?id=11308. Washington State Department of Revenue Website: http://dor.wa.gov/content/taxes/other/tax_rentalcar.aspx

Tobacco Taxes

Description: Tobacco taxes include cigarette taxes and taxes on other tobacco products. The 2006 U.S. federal cigarette excise tax is $.39 per pack. All 50 U.S. states and several U.S. territories have cigarette taxes. Most U.S. states have taxes on other tobacco products as well. Some states earmark a portion of revenues from taxes on cigarettes and/or other tobacco products for environmental purposes. The State of Washington dedicates a portion of its cigarette tax revenues to water quality protection and salmon recovery programs. The State of Idaho uses a portion of its cigarette tax revenues for water quality protection initiatives.

Reference for Further Information: Campaign for Tobacco-Free Kids Website: http://www.tobaccofreekids.org/reports/prices/. National Conference of State Legislatures Website: http://www.ncsl.org/programs/health/Cigarette.htm.
Federation of Tax Administrators Website: http://www.taxadmin.org/fta/rate/tax_stru.html.
The Tax Foundation Website: http://www.taxfoundation.org/blog/topic/103.html.
Washington State Department of Revenue Website:
http://dor.wa.gov/content/taxes/other/tax_cigarette.aspx.
Russell, Betsy Z., "Cigarette tax stays at 57 cents, three Panhandle lawmakers join…..,"
Spokesman Review, The (Spokane), March 30, 2005, available at:
http://www.findarticles.com/p/articles/mi_qn4186/is_20050330/ai_n14589935

Open Space Sales Taxes

Description: Open space sales taxes are levied at the county level to raise money for parks and open space preservation initiatives. For example, in 2004, Adams County, Colorado voters approved a twenty year extension of a one fifth of one percent sales tax that is used to fund the preservation of open space and the creation and maintenance of parks and recreation facilities. In 1990, Sonoma County, California voters approved a quarter cent sales tax that will be extended through 2011 and is used to fund an open space, clean water, and farmland protection measure. Arapahoe County, Colorado electors voted in 2003 to approve a 0.25% open space sales and use tax, effective through 2013, that is used to fund the preservation of open space.

Reference for Further Information: The Trust for Public Land Website:
http://www.tpl.org/tier3_cdl.cfm?content_item_id=4521&folder_id=1365.
Adams County, Colorado Website:
http://webapps.co.adams.co.us/services/department/open_space/resolution.html.
Articles from Petaluma Argus-Courier, a California newspaper:
http://www1.arguscourier.com/apps/pbcs.dll/article?AID=/20061004/NEWS01/61003017.
Arapahoe County, Colorado Website:
http://www.co.arapahoe.co.us/Departments/PW/OpenSpaceProgram/salesandusetax.asp

Taxpayer Contributions Toward Toxic Waste Site Cleanup

Description: In 1995, the polluter pays fees under the U.S. federal Superfund law expired. The polluter pays fees that expired included the crude oil tax, the chemical feedstock tax, and the corporate environmental income tax. These fees could potentially be reinstated by Congress, and groups such as the U.S. Public Interest Groups (PIRGs) are working to make that reinstatement happen. Since the expiration of the Superfund polluter pays fees in 1995, the financial burden for cleaning up toxic waste sites has switched entirely to U.S. taxpayers. Between 1995 and 2006, the burden to taxpayers for cleaning up Superfund sites increased in all U.S. states by hundreds of millions of dollars. For example, the cost to taxpayers for cleaning up Superfund sites in California, a state with 93 Superfund sites, increased from $36,837,101 in 1995 to $461,523,140 in 2004-2005. Taxpayers now pay for all Superfund-led toxic waste site cleanups, spending well over $1 billion annually to protect public health from industrial pollution.

Reference for Further Information: U.S. Public Interest Groups (PIRG) Website: http://www.uspirg.org/home/reports/report-archives/healthy-communities/healthy-communities/on-april-17th-taxpayers-will-pay-to-clean-up-after-polluters-at-toxic-wastes-sites

Gross Receipts Taxes

Description: Gross receipts taxes are levied on the gross amount of money or other compensation, such as barter, that a business receives for its transactions in a given state. The gross taxable amount includes all reimbursed expenses billed to the customer. Examples of these expenses are charges for meals, travel, hotels, shipping, handling, and postage. In some states, gross receipts taxes are assessed in lieu of sales taxes or corporate income taxes. Examples of gross receipts style taxes include Texas's franchise tax on gross receipts of businesses, which is levied at a rate of 0.5% for retailers and 1% for other businesses, New Mexico's gross receipts tax, Kentucky's alternative gross receipts tax, Michigan's Single Business Tax, Washington's Business and Occupations Tax, and Ohio's Corporate Activity Tax. Gross receipts tax revenues from particular businesses could be dedicated to initiatives addressing the environmental impacts created by that business. For example, revenues from the gross receipts of dry cleaning businesses could be used to fund small source air emissions reduction programs.

Reference for Further Information: Tax Foundation Website: http://www.taxfoundation.org/, search the Website on "gross receipts taxes." Contact the taxation and revenue offices for specific states to find out if they levy these taxes. State of New Mexico Taxation & Revenue Website: http://www.state.nm.us/tax/trd_pubs.htm

Corporate Income Taxes

Description: Corporate income taxes, also called corporate franchise taxes, are based upon the net income earned by corporations. They are levied at the state and federal levels. State and federal dedication of corporate income taxes to environmental protection is not common. Still, corporate income tax revenues could be dedicated to finance environmental programs that stem from a particular corporate activity. For example, if two percent of revenue were collected from mining companies, those funds could be earmarked for erosion control, habitat restoration, and other activities that mitigate the environmental impacts of mining. Similarly, revenues from bottling companies could be used to finance state recycling programs. With a relatively broad revenue base, corporate income taxes can be charged at relatively low rates and still generate significant revenues. They can be used to add pollution control to the overall costs of production.

References for Further Information: National Conference of State Legislatures Website:
http://www.ncsl.org/programs/fiscal/fpphqsrs.htm.
Tax Foundation Website: http://www.taxfoundation.org/research/topic/91.html and
http://www.taxfoundation.org/publications/show/1479.html

Estate and Inheritance Taxes

Description: Estate taxes and inheritance taxes are levied on inherited property at or above a specified value. In the United States tax code, estate taxes are paid by the executor of the estate before the heirs receive it, while inheritance taxes are paid by the heirs after ownership of the estate is passed on to them. These taxes provide a broad revenue base. States could earmark a portion of estate and inheritance taxes to general environmental programs. In addition, estate and inheritance taxes can be structured to provide tax relief for property owners making outright land donations and/or placing conservation easements on inheritance land, even during a donor's lifetime. Many states structure estate and inheritance taxes to provide tax relief to donors for land donated to state or local governments or nonprofit land trusts. The donated land can be purchased and managed initially by state or local land trusts, such as The Nature Conservancy, until the appropriate state or local agency assumes responsibility.

Reference for Further Information: Internal Revenue Service Website:
http://www.irs.gov/businesses/small/article/0,,id=98968,00.html.
The Nature Conservancy Website:
http://www.nature.org/wherewework/northamerica/states/maine/preserves/art13333.html.
The Trust for Public Land Website: http://www.tpl.org/, search the Website on "estate tax."

Individual Income Tax Deductions

Description: Individual income taxes, also called personal income taxes, are assessed at the state and federal levels, and sometimes at the county and municipal levels, based on a percentage of income earned by individuals. Some states, such as Maryland, Virginia, and Colorado, allow landowners who donate easements to take deductions on their state income taxes based on the fair market value of the donated easement. The federal government allows individual income tax deductions for land conservation easement donations as well. On August 17, 2006, the President signed into law a substantial expansion of the federal conservation tax incentive for land conservation easement donations. The new federal law raises the deduction a donor can take for donation of a conservation easement from 30% of their adjusted gross income in any given year to 50%. The law also allows qualifying farmers and ranchers to deduct up to 100% of their income.

Reference for Further Information: The Trust for Public Land Website: http://www.tpl.org/tier3_cdl.cfm?content_item_id=1064&folder_id=825. Little Traverse Conservancy Website: http://www.landtrust.org/TaxInfo/NewTaxInfo.htm. Land Trust Alliance Website: http://www.lta.org/conserve/options.htm. Piedmont Environmental Council Website: http://www.pecva.org/conservation/stateincometax.asp. Maryland Department of Natural Resources Website: http://www.dnr.state.md.us/met/sitc.html. Colorado Department of Revenue Website: http://www.revenue.state.co.us/fyi/html/income39.html. See "Conservation Easements" in Section 8 of this Guidebook.

Local Sales Taxes

Description: Local sales taxes are often add-ons to state general sales and use taxes. They may also exist where there is no state sales tax. Depending on state constitutions, statutes, and home rule traditions, most local governments must seek voter approval to levy local sales taxes. State authorization processes vary. States may give approval to all counties or communities, or limit it to specific localities. Local taxes are usually limited to a specified time period, or a dollar collection total, and are dedicated to a specific use. The dedicated revenue stream may be used to back local general obligation or revenue bonds or to pay for a specific environmental protection program directly. The revenues from local taxes are sometimes used to capitalize local revolving funds for environmental protection purposes. Local sales taxes can support a multitude of environmental protection programs. Local sales tax revenues are often dedicated to initiatives such as open space acquisition, wetlands protection, or watershed protection.

Reference for Further Information: National Conference of State Legislatures Website: http://www.ncsl.org/programs/fiscal/fpphqsrs.htm. Tax Foundation Website: http://www.taxfoundation.org/research/topic/9.html. See Department of Revenue Websites for individual states.

Tangible Property Taxes

Description: Tangible property taxes are levied on the estimated or assessed value of items of personal property, such as automobiles and boats, but not land. Such taxes are charged on a recurrent basis, frequently annually or biannually, and sometimes are limited to property worth in excess of a specified dollar value. These taxes are used by state and local governments for a variety of purposes, but they are not typically earmarked for environmental protection purposes. Still, there are many potential environmental protection related uses for these taxes. State and local governments could structure tangible property taxes to mitigate the negative environmental impacts of the use of specific types of tangible property. For example, the revenues generated by a tax on air conditioners could be used for Freon disposal; revenues from a tax on lawnmowers and small engines could be used to fund small source air emissions reduction programs. States could also structure personal property taxes to encourage emissions reduction and/or energy efficiency by discounting tax rates on energy efficient, low emissions vehicles and high-efficiency appliances such as heaters, refrigerators and air conditioners.

Reference for Further Information: National Conference of State Legislatures Website: http://www.ncsl.org/programs/fiscal/fpphqsrs.htm.

Real (Ad Valorem) Property Taxes

Description: Real property taxes, also called ad valorem taxes, are charged to property owners as a percentage of the assessed value of real estate or personal property. They are administered by local governments and require voter approval. Property taxes are an important form of revenue for local governments and they are often used as a funding mechanism for parks and open space measures. There are two main ways localities use property taxes to fund environmental projects. The first is to earmark a specific portion of annual revenues, which is rare. The second is to direct a property tax increase or surcharge, temporary or permanent, to a specific purpose. Revenues from property taxes can go to local trust funds, serve as collateral for general obligation or revenue bonds, and leverage state funds. Property taxes are not well accepted by tax payers compared to other types of taxes, but they provide a steady source of revenue and are less affected by downturns in the economy compared to sales and income taxes; and revenues from them can be accurately predicted.

Reference for Further Information: The Trust for Public Land Website: http://www.tpl.org/tier3_cdl.cfm?content_item_id=1053&folder_id=825. National Conference of State Legislatures Website: http://www.ncsl.org/programs/fiscal/fpphqsrs.htm. Investopedia: http://www.investopedia.com/terms/a/advaloremtax.asp

State Sales and Use Taxes

Description: State sales taxes are a form of excise charged as a percentage of gross retail sales of tangible property. Since state sales taxes only apply to purchases within a state's borders, they are usually accompanied by use taxes that apply to the use, storage, or other consumption within the state of goods or services purchased out of state. The revenue base generated by state sales and use taxes is broad and relatively stable. Some states earmark a specified percentage of their sales and use tax revenues for specific purposes. Revenues from state sales and use taxes could be used to fund many different types of environmental protection initiatives and projects. For example, New Jersey earmarks a percent of its sales taxes for land conservation through its Green Acres Program. It is important to consider, however, that some states may have statutory limitations on general sales tax increases and earmarking.

Reference for Further Information: U.S. Environmental Protection Agency Website:
http://www.epa.gov/owow/nps/MMGI/funding.html#6.
Tax Foundation Website: http://www.taxfoundation.org/research/topic/9.html and
http://www.taxfoundation.org/research/topic/85.html.
New Jersey Green Acres Program Website: http://www.state.nj.us/dep/greenacres/.

Solid Waste Collection Taxes

Description: Many waste management service providers charge solid waste collection taxes to their customers. For example, the State of Washington requires solid waste collection firms within its borders to charge solid waste collection taxes to each of their customers. Washington uses the revenues from these taxes to provide financial assistance to local governments for repair and maintenance of public works projects, such as streets and sewers. The revenues from the solid waste collection taxes go into Washington's Public Works Assistance Account, which is administered by the Washington State Public Works Board. Minnesota is an example of another state that requires waste management service providers within its borders to charge waste collection taxes, called solid waste management taxes, to their customers. Seventy percent of the revenues from Minnesota's solid waste management tax, or at least $33.76 million, are deposited into the state's Solid Waste Fund. The remainder is deposited into Minnesota's general fund with biennial appropriations for county recycling block grants and related solid waste activities.

Reference for Further Information: Washington State Public Works Board Website:
http://www.pwb.wa.gov/. Washington State Department of Revenue Website:
http://dor.wa.gov/content/taxes/other/tax_refuse.aspx. Minnesota Department of Revenue
Website: http://www.taxes.state.mn.us/special/waste/index.shtml.

Severance Taxes

Description: Severance taxes are excises on natural resources extracted or "severed" from the earth. They are quantified based on the level of resource extraction or the market or gross value of the resources extracted or produced. Most U.S. states levy some form of severance taxes. They are usually payable by the severer or producer. In some states payment is made by the first purchaser. Severance taxes include fuel/mineral taxes (based on the volume of coal, gas, oil, or minerals withdrawn), timber taxes (based on the volume of timber logged), and oyster/shellfish taxes (based on the volume or value of shellfish harvested). Several major energy producing states in the U.S. reported a significant rise in 2005 severance tax collections related to the recent rise in energy prices. Severance taxes could be used by any state, and revenues dedicated to activities that mitigate the environmental impacts of natural resources extraction, such as habitat restoration. Severance taxes can yield significant revenues, which could be sufficient to dedicate to capital generation for environmental infrastructure such as wastewater treatment plants.

Reference for Further Information: National Conference of State Legislatures Website: http://www.ncsl.org/programs/fiscal/severtax05.htm.

1. Water and Sewer Capacity Credits
2. Bond Issuance Fees
3. Connection Fees
4. Franchise Fees
5. Septic System Inspection Fees
6. Hunting and Fishing License Fees
7. Aquifer Protection Area Fees
8. Water and Wastewater Utility User Fees
9. Permitting Fees for Projects Affecting Navigable Waters
10. Professional Certification Fees
11. Pay as You Throw Fees
12. State Public Water Supply Withdrawal Fees
13. Tolls
14. Transporter Fees
15. Water Rights Application Fees
16. Septic System and Well Permit Fees
17. Recycling and Disposal Fees
18. Fertilizer and Pesticide Fees
19. National Pollutant Discharge Elimination System Permit Fees
20. Emission Charges
21. Exactions
22. Special Assessments
23. Impact Fees

Water and Sewer Capacity Credits

Description: Water and sewer capacity credits, also called access rights, are charged on a one-time basis to new users requesting access, and old users requiring increases in capacity, to water and sewer facilities. In exchange for payment, applicants are guaranteed future access to a contracted amount of system capacity that has been reserved for their use. This is important because possible sewer moratoriums at a later date could prohibit new residential or commercial development. Many local governments and utility authorities sell water and sewer capacity credits to finance expansion of, or upgrades to, water and sewer systems. Water and sewer capacity credit programs are structured differently in different communities. Columbus, Ohio and the Upper Merion Municipal Utility Authority in Pennsylvania are examples of two jurisdictions that utilize capacity credits charged to customers as a financing mechanism for water and sewer facilities.

Reference for Further Information: University of Maryland Environmental Finance Center Website: http://www.efc.umd.edu/appendixB.html, see "Case: Developer Financing" at that URL. City of Columbus, Ohio Website: http://downtownplan.columbus.gov/skylinedec02.htm.

Bond Issuance Fees

Description: Bond issuance fees above and beyond the standard bond "cost of issuance" fees could be imposed by governments and special authorities on environmental protection related municipal bonds such as infrastructure construction bonds. Cost of issuance fees are assessed as a percentage of total bond value on many different types of municipal bonds. Additional bond issuance fees could be levied by any state or local government or special authority issuing bonds, and the revenues dedicated to a general infrastructure capital account. Fee proceeds might be used to lower specific debt reserve fund requirements, pay for bond insurance or legal fees, or provide low or no interest loans for environmental protection purposes.

Reference for Further Information: See Section 2 of this Guidebook for general information about using bonds to finance environmental protection initiatives. The Bond Market Association, *Fundamentals of Municipal Bonds*, Fifth Edition: New York, NY, 2001, available through the Bond Market Association Website at http://www.bondmarkets.com/story.asp?id=788.
National Council of Health Facilities Finance Authority Bond Terminology page: http://www.nchffa.com/Bond%20Terminology.htm. InvestinginBonds.com Glossary of Bond Terms: http://www.investinginbonds.com/story.asp?id=52#I.

Connection Fees

Description: Connection fees, also called hookup fees, are charged to property owners at the time they connect with existing municipal drinking water and wastewater treatment facilities. Connection fees are generally levied by local governments or county governments. Some local governments charge low or no connection fees, particularly for businesses, essentially subsidizing the costs for drinking water and wastewater treatment and distribution with general revenues. Charging connection fees would provide local governments with a reliable source of revenues to finance drinking water and wastewater treatment plants and allow general revenues to be used for other purposes. Stockton, California and Camden County, New Jersey are examples of jurisdictions charging water and sewer connection fees.

Reference for Further Information: Raftelis Financial Consultants, Inc. Website:
http://www.raftelis.com/default.html, phone # 704-373-1199.
Raftelis, George, *Comprehensive Guide to Water and Wastewater Finance and Pricing*, third edition, CRC Press, 2005; available at http://www.amazon.com.
Stockton, California Website: http://www.stocktongov.com/MUD/General/permits_fees.cfm.
Camden County, NJ Website: http://www.ccmua.org/connfees.html#whatis.

Franchise Fees

Description: Franchise fees can be imposed on any private enterprise that must purchase a franchise to operate a commercial business. In order to become a franchise, a business has to pay a franchise fee. Franchise fees generally start at less than $10,000, and they can exceed $100,000. Mobile and home-based businesses usually pay franchise fees below $10,000. Franchise fees are sometimes charged to utilities and other businesses as a percentage, often around 3%, of their gross. New private businesses will often purchase franchises to market parent goods or services of parent companies. In that scenario, franchise fees are often imposed by state or local governments on the new businesses. Franchise fees levied by state and local governments could be dedicated to environmental protection programs. Some states and localities use revenues from franchise fees to finance parks and recreation facilities. For example, Denver, Colorado uses revenues it receives from franchise fees charged to public utilities to help fund public safety, parks and recreation, public works, and other city services.

Reference for Further Information: AllBusiness.com, Inc. Website:
http://www.allbusiness.com/buying-selling-businesses/franchising-franchise-fee/2182-1.html.
Denvergov.org: http://www.denvergov.org/Mayor/1688press2033.asp.

Septic System Inspection Fees

Description: Septic system inspection fees are the charges for inspections of septic systems carried out by states and counties. Some states and counties require septic system inspections to be done periodically; and they charge septic system fees to finance the inspections. Prospective home buyers routinely have septic system inspections done before they purchase their homes. The State of Arizona mandates septic system inspections and finances them with septic system fees; and Leelanau County, Michigan considered doing the same. Septic system inspection fees capture revenues from households not connected to municipal sewers, but potentially impacting water quality due to septic tank leakage. Septic system inspection fees could be used to finance the creation of septic tank management districts to monitor and prevent spillage. Approximately 25% of the population in North America relies on septic tanks for sewage treatment.

Reference for Further Information: Arizona Department of Environmental Quality Website: http://www.azdeq.gov/environ/water/permits/wastewater.html#onsitefees.
"Septic inspection resurrected," *Leelanau Enterprise*, available at http://www.leelanaunews.com/editorial.php?id=456.

Hunting and Fishing License Fees

Description: Many states in the United States charge fees for the initial awarding and the renewal of hunting and fishing licenses. The revenues from these fees are often used for environmental programs geared towards protecting fish and wild game habitat and for regulation of hunting and fishing. Saltwater fishing license fees are used to raise funds for marine fisheries management programs in many states. Saltwater fishing license fees provide between one and three million dollars annually in smaller states, and up to seventeen million dollars annually in larger states, for marine fisheries management. In a survey of state marine fisheries management agencies conducted by Angling4Oceans, every respondent reported that the existence of saltwater fishing licenses improved their agencies' ability to support angling opportunities because the revenues from license sales were dedicated to resource management. Wisconsin is an example of a state that funds fish and wildlife conservation programs with revenues from hunting and fishing license fees. Seventy-three percent of the budget for Wisconsin's Fish and Wildlife Account comes from fees charged for hunting and fishing licenses and stamps.

Reference for Further Information: U.S. Fish and Wildlife Service Website: http://www.fws.gov/policy/605fw3.html & http://www.fws.gov/policy/605fw2.html.
Wisconsin Department of Natural Resources Website: http://dnr.wi.gov/invest/conservation/money_spent.htm.
Angling 4 Oceans Website: http://www.angling4oceans.org/license.html.

Aquifer Protection Area Fees

Description: Aquifer Protection Area fees are charged for withdrawals of subterranean water and on-site sewage disposal within Aquifer Protection Areas. Aquifer Protection Areas are delineated around wells serving as public water supplies. For example, the State of Washington authorizes counties within its borders to establish Aquifer Protection Areas and charge Aquifer Protection Area fees. The revenues raised with these fees are used to fund initiatives including the construction of wastewater treatment facilities and the preparation of a comprehensive plan to protect, preserve, and rehabilitate subterranean water. Spokane County, Washington charges Aquifer Protection Area fees for water withdrawal and on-site sewage disposal within the Spokane-Rathdrum Aquifer Protection Area. In 2004, a ballot measure to collect the Aquifer Protection Area fees for twenty more years was approved by Spokane County voters.

Reference for Further Information: Washington State Legislature Website:
http://apps.leg.wa.gov/RCW/default.aspx?cite=36.36.
Spokane County Website:
http://www.spokanecounty.org/utilities/faq.asp?faqid=8&section=Aquife.
"Spokane County Voters extend aquifer fees; Aquifer Protection Area," Spokesman Review, Nov. 11, 2004, available at:
http://findarticles.com/p/articles/mi_qn4186/is_20041111/ai_n11705929.

Water and Wastewater Utility User Fees

Description: User fees are the charges to industrial, commercial and residential customers for the use of water and wastewater utility services. These fees are used to finance the utility services. Customers receiving services are connected to central publicly or privately-owned facilities. Water meters and pollutant tracking have led to sophisticated billing procedures and rate structures based on volume and toxicity. Utilities can assess rates to cover their full costs including capital cost recovery ("full cost pricing"), or subsidize the costs of service with general revenues. User fees are limited to localities. A basic issue in rate-setting is the rate base and structure. Ascending block rates are sometimes used for conservation and other purposes.

Reference for Further Information: Raftelis Financial Consultants, Inc. Website:
http://www.raftelis.com/default.html, phone 704-373-1199.
Raftelis, George, *Comprehensive Guide to Water and Wastewater Finance and Pricing*, 3rd edition, CRC Press, 2005; available at http://www.amazon.com.
American Water Works Association, *Principles of Water Rates, Fees, and Charges,* 5th edition, 2000; available at http://www.awwa.org/bookstore/product.cfm?id=30001.
See "U.S. Environmental Protection Agency: Water and Wastewater Pricing Website," a tool in Section 5 of this Guidebook. U.S. Environmental Protection Agency Water and Wastewater Pricing Website: http://www.epa.gov/water/infrastructure/pricing/index.htm.

Permitting Fees for Projects Affecting Navigable Waters

Description: Many states and local governments levy fees on permit applications for proposed projects affecting the course, current, or cross-section of lakes, wetlands, rivers, and streams. The revenues from these fees are often used to cover the processing costs for the permit applications. These permits include federal Army Corps of Engineers permit programs under Section 404 of the Clean Water Act and Section 10 of the Rivers and Harbors Act. They also include state and local permit programs, such as the Minnesota Department of Natural Resources' Public Waters Work Permit Program, Wetland Conservation Act permits administered by local governments in Minnesota, and permits under the Virginia Water Protection Permit Program. These types of permits help to protect wetlands and other water bodies from the impacts of development because specific environmental protection requirements must be met for the permits to be approved.

Reference for Further Information:
U.S. Army Corps of Engineers Website: http://www.usace.army.mil/cw/cecwo/reg/oceover.htm.
Minnesota Department of Natural Resources Website:
http://www.dnr.state.mn.us/waters/watermgmt_section/pwpermits/applications.html.
Virginia Department of Environmental Quality Website:
http://www.deq.virginia.gov/wetlands/permitfees.html.

Professional Certification Fees

Description: Professional certification fees are charged to companies or individuals for the privilege of obtaining specific professional certifications. Fees are charged for certifications in disciplines including: construction management, wastewater system operation, water supply and wastewater engineering, air pollution control, general environmental engineering, hazardous waste management, industrial hygiene, radiation protection, solid waste management, and disposal and reuse of construction and demolition materials. The revenues raised with these fees can be used to fund environmental protection programs and training for professionals in environmental protection related industries.

Reference for Further Information:
American Academy of Environmental Engineers Website:
http://www.aaee.net/Website/WhyCertified.htm.
Water Environment Federation Website:
http://www.wef.org/ConferencesTraining/TrainingProfessionalDevelopment/WastewaterOperatorsCertification/.
Solid Waste Association of North America Website:
http://www.swana.org/www/EDUCATE/CERTIFICATION/ConstructionDemolitionMaterials/tabid/94/Default.aspx.
Construction Institute Website:
http://www.construction.org/index.php?src=gendocs&link=ProfessionalCertification&submenu=Education.

Pay as You Throw Fees

Description: In communities with pay-as-you-throw (PayT) programs, residents are charged for the collection of municipal solid waste (ordinary household trash) based on the amount they throw away. This creates an economic incentive for residents to recycle more and generate less waste. PayT programs are a form of unit pricing, also called variable rate pricing. Most communities with PayT programs charge residents a fee for each bag or can of waste they dispose of. In some communities, residents are billed based on the weight of their trash. The U.S. Environmental Protection Agency supports the pay-as-you-throw approach to solid waste management because it encourages environmental sustainability, economic sustainability, and equity. Communities with PayT programs in place have reported significant increases in recycling and reductions in waste, due mainly to the waste reduction incentive created by PayT. This suggests that PayT Programs are environmentally sustainable. Well-designed PayT programs are economically sound because they generate the revenues communities need to cover their solid waste costs. PayT programs are equitable because residents pay less if they generate less waste.

Reference for Further Information: U.S. Environmental Protection Agency Website:
http://www.epa.gov/epaoswer/non-hw/payt/intro.htm.
Eco-cycle Website: http://www.ecocycle.org/zero/pay_throw.cfm.
GreenWorks Website: http://www.greenworks.tv/wastemanagement/payt.htm.
WasteAge Website: http://wasteage.com/mag/waste_payasyouthrow/index.html.

State Public Water Supply Withdrawal Fees

Description: State public water supply withdrawal fees are charged for permits for large quantity water withdrawals, generally 10,000 gallons per day or more. Virginia and Michigan are examples of states that charge these water withdrawal fees. Virginia charges public water supply withdrawal fees to subdivisions, public water systems, stores, and other entities for the right to withdraw water in quantities ranging from 10,000 gallons to over 100,000 gallons per day. "Large quantity withdrawals" are defined by the State of Michigan as water withdrawals of greater than 100,000 gallons per day averaged over a consecutive 30-day period. Michigan requires permits for certain new or increased large quantity water withdrawals, including water withdrawals from a Great Lake and water withdrawals from inland lakes or streams; and the state charges a $2,000 fee for these permits. The demand for public water, particularly by industry, is relatively inelastic, resulting in stable and predictable revenues from these fees.

Reference for Further Information: Virginia Department of Environmental Quality Website:
http://www.deq.state.va.us/gwpermitting/pnarchive.html.
Announcement of new water withdrawal law for Michigan:
http://www.deq.state.mi.us/documents/deq-wd-withdrawallaw-summary.pdf.

Tolls

Description: Tolls are fees charged for vehicle passage on thruways, highways, roads and bridges to offset expenses for construction, operation and maintenance. States charge tolls to raise money for transportation budgets. The Transportation Act for the 21st Century (TEA-21), signed into law in 1998, has provisions in it permitting states to use specified toll revenue expenditures as credits towards the non-federal matching share of grants for public transit programs authorized by the Act.

In addition, tolls can be used to raise the funds necessary to add features to roads that minimize environmental impacts or encourage environmentally friendly activities such as walking or biking. For example, Florida's Suncoast Parkway is a toll road that is designed for minimum impact on the environment and maximum use by nonmotorized traffic. It has features including a bike trail alongside it and wildlife crossings which may not have been possible to fund without the use of tolls.

Reference for Further Information: See "Transportation Equity Act for the 21st Century" in Section 2c of this Guidebook. See the U.S. Department of Transportation Website at http://www.fhwa.dot.gov/Tea21/tollcred.htm for information on TEA-21, and at http://www.tfhrc.gov/pubrds/03sep/01.htm for information on Florida's Suncoast Parkway.

Transporter Fees

Description: Transporter fees are charged by states to individuals and corporations for the right to transport solid waste, hazardous waste, petroleum products, and radioactive waste. Revenues from these fees are used by states to pay the costs of hazardous waste monitoring and spill response and other environmental protection initiatives. For example, hazardous waste transporters pay a fee based on the quantity of waste they collect and/or deliver in Massachusetts. The revenues raised with this fee are used to fund the cleanup of hazardous waste sites and spills in Massachusetts. Also, transporters pay a fee for hazardous waste transportation in Pennsylvania. Revenues raised with this fee are deposited into Pennsylvania's Hazardous Sites Cleanup Fund. The State of Connecticut charges application fees for permits for the transportation of waste oil, petroleum and chemical liquids, hazardous waste, and biomedical waste. The revenues from these fees go into the state's Environmental Quality Fund, which can be used for any environmental protection purpose authorized by the fund's commissioner.

Reference for Further Information: Massachusetts Department of Environmental Protection Website: http://www.mass.gov/dep/recycle/approvals/transinf.htm.
Pennsylvania Code: http://www.pacode.com/secure/data/025/chapter263a/subchapBtoc.html.
Connecticut Department of Environmental Protection Website: http://dep.state.ct.us/pao/weedfact/wastrans.htm.

Water Rights Application Fees

Description: State water rights application fees are imposed on applicants for water appropriations. A water appropriation is an authorization granted by a state to make private, beneficial use of the state's water resources. Approved appropriations exist in the form of permits authorizing the use of either ground water or surface water. Fees are charged for applications for new appropriations; and they may be charged on a recurrent basis as well. Most western states charge water rights application fees. Revenues from water rights application fees are used to cover the costs states incur when they carry out various activities necessary to process the applications for new appropriations. These activities include site investigation, performance of environmental and hydro geologic analyses, investigation into whether the water is available or would impair other water users, and preparation of a report with the investigators' findings and a recommendation of whether or not to approve the application.

Reference for Further Information: Bureau of Land Management Website: http://www.blm.gov/nstc/WaterLaws/abstract1.html. Washington State Department of Ecology Fact Sheet: http://www.ecy.wa.gov/pubs/0511016.pdf. South Dakota Department of Environment & Natural Resources Website: http://www.state.sd.us/denr/des/waterrights/faqwr.htm.

Septic System and Well Permit Fees

Description: Many states, counties, and municipalities charge fees for the issuance of construction and use permits for septic systems and wells. Septic system permit fees are charged for permits required for new septic system installations, review of abandoned septic systems, septic system operators' licenses, and septic system pumping. Well permit fees are charged for permits required for the drilling of residential wells, well repairs and modifications, and inspections of wells, and for permits required for licensing companies and individuals who construct wells. Randolph, New Jersey and Monroe County, Michigan are examples of jurisdictions that charges fees for septic system and well permits. Delaware is an example of a state that charges fees for well permits. These fees help to cover the administrative costs of issuing these permits. The permits help to ensure that septic systems and wells are properly constructed and maintained in accordance with environmental laws and health regulations.

Reference for Reference for Further Information: Town of Randolph, New Jersey Website: http://www.randolphnj.org/townhall/septic_wells_permits/.
Monroe County, Michigan Website: http://www.co.monroe.mi.us/monroe/default.aspx?PageId=256.
Delaware Division of Water Resources Website: http://www.dnrec.state.de.us/water2000/Sections/WatSupp/WellPermits/WSSWellPermits.htm.

Recycling and Disposal Fees

Description: Recycling and disposal fees are levied on hard-to-dispose items that contribute heavily to solid waste disposal problems, such as tires and lead acid batteries. Florida, Maine, South Carolina, Texas, and Wisconsin charge fees at the point of sale on lead acid batteries. The revenues raised with fees on lead acid batteries are used to fund the proper disposal of the batteries and solid waste disposal programs in general. Alaska, Maryland, Nevada, Maine, and Florida are examples of U.S. states that charge fees on purchases of new tires. The revenues raised with these fees on new tires are used to fund recycling of the tires.

Reference for Further Information: Battery Council International Website: http://www.batterycouncil.org/states.html.
U.S. Environmental Protection Agency Website: http://www.epa.gov/jtr/state/funding.htm.
Comptroller of Maryland Website: http://business.marylandtaxes.com/faq/tirefaq.asp.
Alaska Tax Division Website: http://www.tax.state.ak.us/programs/tire/faq.asp.
Florida Department of Revenue Website: http://dor.myflorida.com/dor/taxes/vehicle_fees.html.
Federation of Tax Administrators Website, with links to the revenue departments of all 50 U.S. States and Puerto Rico: http://www.taxadmin.org/fta/link/.

Fertilizer and Pesticide Fees

Description: Fertilizer and pesticide fees include dealer license fees, assessment and inspection fees, and registration fees. States often use these fees to raise money for agriculture related environmental protection initiatives. Iowa, Montana, and Nebraska are examples of U.S. states that have enacted fertilizer and/or pesticide fees. The State of Iowa charges pesticide fees authorized by the 1987 Groundwater Protection Act. A portion of the revenues raised with these pesticide fees is placed in the agriculture management account of Iowa's groundwater protection fund. Montana charges pesticide and fertilizer registration fees and uses the revenues it raises with the fees to fund groundwater quality monitoring work. Nebraska charges fertilizer inspection fees and pesticide registration fees and dedicates the revenues raised with the fees to programs for regulating fertilizers and pesticides.

References for Further Information: The New Rules Project Website:
http://www.newrules.org/environment/iaground.html. Leopold Center For Sustainable Agriculture Website: http://www.leopold.iastate.edu/about/igpa.htm.
City of Iowa City Website: http://www.icgov.org/water/groundwateract.htm.
State of Montana Website: http://www.agr.state.mt.us/licensing/fees.asp#feed1.
Nebraska Department of Agriculture Website:
http://www.agr.ne.gov/regulate/bpi/pes/actbm.htm#2634.

National Pollutant Discharge Elimination System Permit Fees

Description: The National Pollutant Discharge Elimination System (NPDES) Permit Fees legislation, passed in 2004, amends the Natural Resources and Environmental Protection Act (NREPA) and authorizes state environmental departments to collect NPDES Permit Fees consisting of annual permit fees and permit application fees. The NPDES Permit Fees are used as a means to obtain the appropriate funding to effectively operate the NPDES program. In addition, the Permit Fee Incentive for Clean Water Act (CWA) Section 106 Grants, an amendment proposed in 2006, would be a means for the U.S. Environmental Protection Agency (EPA) to provide a financial incentive for states to utilize a fee program that is adequate to provide sufficient funding when implementing an authorized NPDES permit program. EPA proposed this CWA Section 106 amendment, through an official publication in the Federal Register in 2007, to provide the Agency with the authority to allot a permit fee incentive amount. This action would not be effective prior to fiscal year 2008.

Reference for Further Information: U.S. Environmental Protection Agency Website: http://cfpub.epa.gov/npdes/ and http://www.epa.gov/owm/cwfinance/npdes-permit-fee.htm. Michigan Department of Environmental Quality Website: http://www.michigan.gov/deq/0,1607,7-135-3313_3682_3713-90130--,00.html. Illinois Environmental Protection Agency Website: http://www.epa.state.il.us/fees/npdes.html.

Emission Charges

Description: Emission charges, also called emission fees, are levied based on the volume and toxicity of pollutants emitted into the atmosphere by industries, municipal facilities such as power plants, and motor vehicles. States charge a type of emission-based permit fee under Title V of the Clean Air Act, which requires states to charge permitted sources the equivalent of $25 per ton of regulated pollutants emitted. Since the purpose of this requirement is to help states recover the full cost of permit issuance, these fees are also called permit fees. In addition, Distance-Based Emission Fees, which are sometimes charged at the state level, are mileage-based charges that reflect a vehicle's emissions rate. Vehicles with lower emissions pay lower Distance-Based Emission Fees than vehicles with higher emissions. A major advantage of emissions charges and fees is that they create a strong incentive for those paying them to innovate and thus discover less expensive ways of reducing pollution. Emissions charges attack the pollution problem at its source by putting a price on the "right to pollute." The "right to pollute" has historically been free and thus overused.

Reference for Further Information: Environmental Economics, Third Edition, Chapter 12, McGraw-Hill: 2001, available at http://www.mhhe.com/economics/fieldee/about/toc.mhtml. Online TDM Encyclopedia: http://www.vtpi.org/tdm/tdm59.htm#_Toc29092444. Connecticut Department of Environmental Protection Website: http://dep.state.ct.us/pao/airfact/titlev.htm

Exactions

Description: Exactions, also called proffers, are conditions or financial obligations imposed on developers to aid local governments in providing public services needed to support new developments. They are administered by local governments. Exactions can take a number of different forms. They can include financing of existing infrastructure facilities or infrastructure improvements, donations of in-kind services, and donations of land, water and sewer lines, and road and parking facilities. Exactions can also take the form of impact fees paid in lieu of the types of donations described above. Exactions have the benefit of allowing more flexibility than impact fees because they are not required to be financial contributions. They may be offered voluntarily by developers; and local governments often negotiate them with each developer. Most localities use exactions in some form. Some localities assign building permits competitively based on the level of exactions offered by different developers.

Reference for Further Information: See "Impact Fees" in this section of the Guidebook.
Facsnet: http://www.facsnet.org/tools/env_luse/nat9exactions.php3.
Public Policy Institute of California Website: http://www.ppic.org/main/publication.asp?i=106.
County of Albemarle, Virginia Website:
http://www.albemarle.org/department.asp?department=planning&relpage=7084.

Special Assessments

Description: Special assessments are recurrent surcharges levied by local jurisdictions on sub-groups of the population. Some localities levy them in the form of taxes; others levy them in the form of fees. The sub-group paying the recurrent charges receives benefits from an environmental service or improvement not enjoyed by others in the area. For example, if a community wants to finance wastewater treatment plant improvements that contribute to lake cleanup, residents with waterfront property could be charged a special assessment. Special assessments are generally charged by local governments and authorized by local ordinance. They are often barred by constitution from use by states. Special assessments are used to fund water works systems, sanitary sewer systems, installation or repair of water and sewer service lines, flood protection projects, and other purposes. Minneapolis, Minnesota; Fargo, North Dakota; and Manhattan, Kansas are examples of cities charging special assessments.

Reference for Further Information: TaxProf Blog Website:
http://taxprof.typepad.com/taxprof_blog/2007/01/baker_on_using_.html.
Minneapolis, Minnesota Website: http://www.ci.minneapolis.mn.us/special-assessments/.
Fargo, North Dakota Website: http://www.cityoffargo.com/Residential/SpecialAssessments/.
Manhattan, Kansas Website: http://www.ci.manhattan.ks.us/index.asp?NID=285.

Impact Fees

Description: Impact fees are frequently assessed on the construction of new buildings. Local governments and county governments levy impact fees. The revenues from impact fees are used to pay for improvements to services and amenities necessary to serve the occupants of the new buildings, including expansions of police and fire stations, sewer and water supply systems, parks, libraries, and schools, and the building of new roads. In addition, impact fees are frequently assessed based on the projected environmental impacts of a construction project; and the revenues from the fees are used to mitigate the project's environmental impacts. Environmental impacts of construction projects include erosion of land where vegetation has been removed and additional storm water runoff due to the installation of pavement and other impervious surfaces. Collier County, Florida and Fremont, California are examples of jurisdictions that charge impact fees on new construction projects.

Reference for Further Information: Wisconsin Realtors Association Website: http://www.wra.org/Government/Land_Use/impact_fees/default.htm.
Collier County, Florida Website: http://www.colliergov.net/Index.aspx?page=1538. Fremont, California Website: http://www.ci.fremont.ca.us/Construction/Fees/DevImpactFee1.htm.

This section presents the three major ways in which governments and the private sector acquire capital to invest in environmental protection initiatives including pollution prevention: bonds, loans, and grants. Bonds and loans entail repayments of principal and interest, although interest rates may be governmentally subsidized. Grants are awarded for specifically designated purposes and do not require repayment. These three different means for acquiring capital are presented in subsections 2a (bonds), 2b (loans), and 2c (grants). Each means for acquiring capital covered in this section serves a distinct purpose and has certain limitations.

A bond is a written promise to repay borrowed money on a definite schedule, and usually at a fixed rate of interest, for the life of the bond. Some types of bonds are tax exempt. Bonds represent a large source of capital, but can be a complex and more expensive way to borrow. The high expense results from the legal and other fees and administrative time required for issuing bonds. In some cases voter approval is required for issuing bonds.

Loans typically involve fewer and lower transaction costs than bonds. Interest rates on government loans may be subsidized, particularly for small communities. Like grants, government loans are made with very specific goals in mind, are often accompanied by specific mandates, and are limited by legislatively appropriated dollar amounts. Most government loans have complicated application procedures and deadlines. Commercial loans are more flexible than government loans, but are typically more expensive for public and private borrowers.

Grants are generally regarded as more desirable than loans and bonds. However, since grants are designed by the awarding agency or organization to meet certain, often specific, goals, they may carry additional mandates as compared to loans and bonds; and those mandates may be costly to meet. Grants tend to have difficult application procedures and deadlines. Grant money often comes from tax dollars. The redistribution of federal and state tax revenues to some communities and not others can be controversial. Many grant programs must be approved annually by legislative bodies.

There are, of course, many additional types of bonds, loans, and grants that are not covered in this section. Additional grants and loans are described in Section 6, which is the "State and Local Tools" section, and in other sections of the Guidebook as well. Information on additional loans and grants not listed in the Guidebook can be found through many sources, including the Catalogue of Federal Domestic Assistance (CFDA) and Grants.gov. The CFDA is a searchable database with detailed information on all federal financial assistance programs. Grants.gov is a searchable database with comprehensive information on over 1,000 grant programs offered by all federal grant making agencies.

All of the means for acquiring capital covered in this section (bonds, loans, and grants) are available for use alone or in combination to fund specialized environmental protection initiatives. An example of combining different financing mechanisms is when states utilize loans or bonds to acquire dollars needed for matching fund requirements on grants. When carefully matched with the recipient's needs these tools, whether used alone or in combination, can be very effective in the future success of environmental protection initiatives.

1. Advance Refunding Bonds
2. Anticipation Notes
3. Appropriation-Backed Bonds
4. Asset-Backed Securities
5. Capital Appreciation and Zero Coupon Bonds
6. Certificates of Participation
7. Derivatives
8. Double-Barrel Bonds
9. General Obligation Bonds
10. Mini/Baby Bonds
11. Moral Obligation Bonds
12. Private Activity Bonds
13. Revenue Bonds
14. Short-Term Municipal Bonds
15. Special Assessment Bonds
16. Special Tax Bonds
17. State Revolving Fund (SRF) Revenue Bonds
18. Tax Increment Bonds

Advance Refunding Bonds

Description: Advance refunding is the issuance of a new bond to pay off another outstanding bond prior to the date on which the earlier bond can be redeemed or paid. Advance refunding is undertaken primarily to adjust outstanding debt to current interest rates, and/or to alter debt reserve requirements. The proceeds from the sale of the refunding bonds are used to buy taxable government securities, which are deposited in an escrow account. The escrow account is structured so that the principal and interest earned on the securities are enough to pay all principal, interest, and call premium, if any, on the outstanding bonds up to and including the call date. The 1986 Tax Reform Act limits each governmental activity bond issue to one advance refunding if the original issue was after 1985. Thus, bond leveraged State Revolving Funds are limited in their use of advance refunding.

Reference for Further Information: Graham, William; Shinn, P.; Petersen, J., *State Revolving Funds Under Tax Reform,* Council of Infrastructure Financing Authorities (CIFA) Monograph No. 2, Washington, D.C.: June 1989, to order contact CIFA at cifa@navigantconsulting.com. CIFA Website: http://www.cifanet.org/publications.html.
WM Financial Strategies Website: http://www.munibondadvisor.com/refunding.htm.
Investopedia.com Website: http://www.investopedia.com/terms/a/advancedrefunding.asp.

Anticipation Notes

Description: Anticipation notes are short-term bond instruments repaid with anticipated revenues from various sources. They can be used to acquire immediate capital when other funding sources are delayed or not yet identified. For example, if a city anticipated a future federal grant for a project, the government might issue a revenue anticipation note to meet interim construction costs. State and local governments widely use anticipation notes. There are four primary types of anticipation notes: 1.) *Tax Anticipation Notes* (TANs), which are short-term, tax-exempt notes issued in anticipation of tax receipts and paid from those receipts; 2.) *Revenue Anticipation Notes (RANs),* which are issued in anticipation of other sources of future revenues, such as federal or state aid; 3.) *Bond Anticipation Notes (BANs),* which are designed to provide financing until a future bond offering is made, and 5.) *General Obligation (GO) notes,* which are not backed by any particular revenue source, but by the full faith and credit of the issuing government. Interest rates for anticipation notes are typically higher than those for longer-term securities.

Reference for Further Information:
First Public, LLC Website: http://www.firstpublic.com/cities/financing/tan.shtml.
Investopedia.com Website: http://www.investopedia.com/terms/r/ran.asp.

Appropriation-Backed Bonds

Description: Appropriation-backed bonds are state special obligation bonds using a pledge of future state direct appropriations, typically annual appropriations, as the form of pay back to the bondholders. Such bonds may be either tax-exempt or taxable. State bond issuance is authorized by state legislatures, and the issuing authority may enter into a service contract or lease arrangement with the state or state agency undertaking the activity being financed. These bonds can be useful as a financing device to cover special needs which may fall outside of the normal budgeting cycle of state legislatures. Appropriation-backed bonds have been challenged legally in a number of states, on the grounds that legislative appropriation of funds does not constitute adequate assurance for the bondholders. This has made many states cautious about using them. In some states, use of such bonds is prohibited by the state constitution.

Reference for Further Information: The Bond Market Association, *Fundamentals of Municipal Bonds*, Fifth Edition: New York, NY, 2001, available through the Bond Market Association Website at http://www.bondmarkets.com/story.asp?id=788.
New York State Citizen's Guide: http://www.budget.state.ny.us/citizen/financial/capital.html.

Asset-Backed Securities

Description: An asset-backed security is a type of bond that is backed by a pool of financial assets. The financial assets in the pools backing these securities include credit card debt, accounts receivables, and mortgages. The mortgage-backed security market is so large, however, that it is often seen as separate from other asset-backed securities. For asset-backed securities, assets are pooled to make otherwise minor and uneconomical investments worthwhile, while reducing risk by diversifying the underlying assets. When a large portfolio of liquid assets is pooled together; the assets can be converted into instruments that may be offered and sold freely in the capital markets. A significant advantage of asset-backed securities is that they bring together a pool of assets that otherwise could not be traded easily in their existing form. That advantage could help small communities to use asset-backed securities for the purpose of funding environmental protection initiatives.

Reference for Further Information: Hayre, Lakhbir, *Solomon Smith Barney Guide to Mortgage-Backed and Asset-Backed Securities:* John Wiley & Sons, Inc., 2001, available at www.amazon.com.

Capital Appreciation and Zero Coupon Bonds

Description: Capital appreciation bonds (CABs) and zero coupon bonds (zeros) are used in the issuance of state and local general obligation and revenue-backed debts. CABs and zeros provide investors a guaranteed reinvestment rate, so they are most attractive to investors when interest rates are expected to fall. They are original issue discount bonds that are sold at face value (par). The issuer is not required to make periodic interest payments on CABs and zeros. Instead, the interest component is held by the issuer and compounded at a stated rate so the investor receives a lump sum multiple of the principal and interest when the bond matures. CABs and zeros are sold at deep discount from their face value. At maturity date, the security is redeemed at face value. The investor receives a rate of return based on the appreciation from the discounted price to the full face value. CABs are different from zeros in that the investment return on CABs is in the form of compounded interest rather than accreted original issue discount. CABs result in more bond proceeds for the same use of debt capacity (total par value) than do zero coupon bonds.

Reference for Further Information: Springsteel, Ian, "Who Needs Equity?: zero-coupon convertible bonds," *CFO: Magazine for Senior Financial Executives,* July 2001, available at http://www.findarticles.com/p/articles/mi_m3870/is_7_17/ai_77290112. Investorguide.com Website: http://www.investorguide.com/igup1-advanced-bond-concepts.htm.

Certificates of Participation

Description: Certificates of participation (COPs) are financial instruments used to finance capital projects. COPs are backed by the leasing of real property and physical assets, such as wastewater plants or equipment. The assets are held by a trustee, and the certificate issuer pays yearly lease payments to the certificate holders until the debt is repaid. If the certificate issuer defaults on the lease payments, the trustee is responsible for selling the physical assets and using the proceeds to reimburse the certificate holders. Certificates of participation can only be issued to finance capital projects where a real asset exists that is suitable as collateral, and only in jurisdictions where local authorities are allowed to negotiate long-term leases. COPs are similar to mortgage bonds and asset-backed bonds, but are not legally classified as such, so state and local governments can issue them without voter approval and without affecting their overall bonding capacity. Certificates of participation do not count against debt capacity limits. COP payments to private investors are tax-exempt. This financing mechanism is used in more than half of U.S. states.

Reference for Further Information: O'Meara, Kelly P., "Creative financing: dozens of municipal projects in Los Angeles County have been financed using bondlike instruments called COPs, which critics charge have allowed officials to enter into.....," *Insight on the News,* April 15, 2002, available at http://www.findarticles.com/p/articles/mi_m1571/is_13_18/ai_84804831. Orrick Website: http://www.orrick.com/practices/public_finance/leaseRevenue.asp.

Derivatives

Description: A derivative is a financial instrument which derives its value from a specific, underlying market or index. From an accounting perspective, a derivative is defined as having two characteristics: 1.) the holder has the right to participate in some or all of the price change experienced by the underlying market or index, and 2.) the instrument's value at maturity can be settled in cash as opposed to taking ownership of the underlying market or index. For example, a bond paying an interest rate based on changes in the stock market may be called a derivative because its value changes in response to a market. The market may be measured by an index such as Standard and Poor's 500. Many different financial instruments; including swaps, caps, options, puts, calls, and collars; are classified generically as derivative products. These financial instruments all derive their value from the performance of specific indices or cash markets.

Reference for Further Information: The Bond Market Association; Temel, Judy W.; *The Fundamentals of Municipal Bonds*, 5th ed.; John Wiley & Sons, Inc.; 2001, available at http://www.amazon.com.

Double-Barrel Bonds

Description: A double-barrel bond is a municipal revenue bond secured by a pledge of two or more sources of payments, typically a user fee and, secondarily, by the credit of the issuing government through *ad valorem* taxes, which are taxes based on the assessed value of property. State and local governments use double-barrel bonds to finance environmental improvements, including renovation of wastewater treatment plants, construction of drinking water utilities, and creation of stormwater management districts. The revenue stream pledge may be in the form of multiple taxes, such as the real estate transfer tax or special assessment taxes. Double-barrel bonds can provide cheaper capital than conventional revenue bonds for projects that generate revenue. They are a good means for states or localities, particularly those with low credit ratings or low debt capacity, to obtain lower interest rates on bond issues compared to conventional revenue bonds.

Reference for Further Information: "General Obligation Bonds" in this section of the Guidebook. The Bond Market Association; Temel, Judy W.; *The Fundamentals of Municipal Bonds*, 5th ed., John Wiley & Sons, Inc.; 2001, available at http://www.amazon.com. Answers.com: http://www.answers.com/topic/municipal-bond.

General Obligation Bonds

Description: General Obligation (GO) bonds are backed with the guarantee that the issuing government will use its taxing power to repay them. GO bonds are regarded as safer than bonds backed by a single revenue source, and generally command lower interest rates and lower reserve fund requirements. There are two primary types of GO bonds: unlimited *ad valorem* tax debt and limited *ad valorem* tax debt. *Ad valorem* taxes are based on the assessed value of property. Unlimited *ad valorem* tax debt occurs when the government pledges its full faith and credit with no limitations on possible property tax rates. Limited *ad valorem* tax debt occurs when the government pledges its full faith and credit, but with a cap or restriction on possible property tax rates. Occasionally, a GO bond may be backed by a specific revenue source. State and local governments use GO bonds to finance capital projects for environmental protection initiatives such as lands purchases. State referendum environmental bonds, which are often very large, are GO bonds paid for by a variety of sources of revenue including appropriations. GO bonds are suitable for financing projects that require large amounts of capital up-front. Voter approval is frequently required for GO bonds.

Reference for Further Information: The Bond Market Association; Temel, Judy W.; *The Fundamentals of Municipal Bonds*, 5th ed., John Wiley & Sons, Inc.; 2001, available at http://www.amazon.com. Answers.com: http://www.answers.com/topic/municipal-bond. Municipal Bond Information Guide: http://www.southwest.msus.edu/RDIC/gobond.html.

Mini/Baby Bonds

Description: Mini Bonds are characterized by direct marketing from issuers to investors. They are also called baby bonds because of their relatively low face or par values which are generally below $1000. Modeled after federal savings bonds, they bring segments of the bond market within reach of small investors and open a source of funds to issuers without access to the large institutional market. Mini bonds have characteristics designed for investors with various objectives. For example, some are structured as capital appreciation bonds so investors do not have to reinvest periodic interest earnings (see "Capital Appreciation and Zero Coupon Bonds" in this section of the Guidebook). Mini bonds have been used extensively at the state and local levels since the 1970's. They could be used to finance relatively small, targeted environmental investments, such as nonpoint source pollution control measures and stream restoration.

Reference for Further Information: Melvin, Sean P., "Itsy-bitsy bonds: Raising money: No more waiting around till your company hits the multimillion-dollar mark- now bonds are for businesses of all sizes," *Entrepreneur,* January 2002, available at http://www.findarticles.com/p/articles/mi_m0DTI/is_1_30/ai_83790627.
Business Editors, "Lower Colorado River Authority-TX-Rev Mini-Bonds RTD 'AAA,'" Business Wire, New York, Feb. 3, 2000, available at http://www.findarticles.com/p/articles/mi_m0EIN/is_2000_Feb_3/ai_59169468.

Moral Obligation Bonds

Description: A moral obligation bond is a bond secured with revenues from a financed project, as well as a non-binding pledge that any deficiency in pledged revenues will be reported to the state legislature, which may appropriate state monies to make up the shortfall. Under most state laws, if a draw down of the bond's debt reserve occurs, the bond trustee must report the amount used to the governor and the state legislature. The state legislature is then authorized to appropriate the requested amount to repay the bondholders, although there is no legally enforceable obligation to do so. New York, the first state to issue moral obligation bonds, used the bonds to finance a housing authority. Moral obligation bonds can be used to acquire project capital at lower rates than revenue bonds. They generally do not count against debt issuance limitations. Moral obligation bonds can obtain interest rates almost as low as general obligation bonds because they are backed by the pledge of repayment.

Reference for Further Information: Council of Development Finance Agencies Website: http://www.cdfa.net/cdfa/cdfaweb.nsf/pages/sep2003tlc.html. The Bond Market Association; Temel, Judy W.; *The Fundamentals of Municipal Bonds*, 5th ed., John Wiley & Sons, Inc.; 2001, available at http://www.amazon.com.
Investopedia.com: http://www.investopedia.com/terms/m/moralobligationbond.asp.

Private Activity Bonds

Description: "Private activity" or "exempt" is a term used to describe industrial development bonds and other similar types of bonds which meet one of a number of tests under federal tax law measuring private involvement in a bond financing. The most commonly used definition includes bonds which meet both the private business use test and the private payment definition. The private business use test is met when no more than ten percent of bond proceeds are used by an entity other than a state or local government unit. The private payment test is satisfied when no more than ten percent of debt service on the bonds is directly or indirectly paid or secured by a private entity. Most of these restrictions fall under the 1986 Tax Reform Act. State and local private activity bonds may be issued on a tax-exempt basis for specifically identified purposes if a myriad of specific rules are satisfied. Although interest on such bonds is exempt from the regular income tax, interest on the bonds (other than for bonds issued for 501(c)(3) charitable organizations) is an item of "tax preference" for purposes of the alternative minimum tax. Private Activity Bonds are used to fund facilities for the furnishing of water and sewage, solid waste disposal facilities, and qualified educational facilities.

Reference for Further Information: The Bond Market Association; Temel, Judy W.; *The Fundamentals of Municipal Bonds*, 5th ed., John Wiley & Sons, Inc.; 2001, available at http://www.amazon.com. Internal Revenue Service, Tax Exempt and Government Entities, "Tax Exempt Private Activity Bonds," 2004, available at http://www.irs.gov/pub/irs-pdf/p4078.pdf.

Revenue Bonds

Description: "Revenue bond" is a broad term used to describe bonds on which the debt service is payable mainly from revenue generated through the operation of the project being financed, or from other non-property tax sources. They may be issued by state and local governments, or by an authority, commission, special district, or other unit created by a legislative body for the purpose of issuing bonds for facility construction. Revenue bonds now account for the clear majority of municipal bonds used to finance water, sewer, and solid waste infrastructure in the United States. Revenue bonds are usually tax-exempt. Bond interest rates may be higher for revenue bonds compared to general obligation bonds, and even higher for taxable revenue bonds. Revenue bonds do not count against debt ceilings, but the national rating agencies take them into account in financial capability analyses. State Revolving Fund (SRF) bonds, private-activity industrial development bonds, and mortgage lease-backed bonds are examples of revenue bonds.

Reference for Further Information: The Bond Market Association; Temel, Judy W.; *The Fundamentals of Municipal Bonds*, 5th ed.; John Wiley & Sons, Inc.; 2001, available at http://www.amazon.com.
Municipal Bond Information Guide: http://www.southwest.msus.edu/RDIC/revbond.html.
State of Montana Website: http://www.mtfinanceonline.com/typesofbonds.asp.

Short-Term Municipal Bonds

Description: Traditionally, the phrase "short-term municipals" has meant short-term municipal bonds and short-term securities known as notes. There are two main types of notes, anticipation notes and general obligation notes. All of these instruments generally have maturities ranging from a few months to a few years, have fixed interest rates, and are issued in anticipation of a bond issue, grant proceeds, or tax collections. In the 1980s a new, broader class of "short-term municipals" was developed. These new "short-term municipals" are known as demand obligations or variable rate demand obligations. They are long-term bonds with yields determined as if they were short-term notes. The bond holders can demand purchase of their bonds at par (the principal due at maturity), plus accrued interest. The interest rates vary at predetermined intervals, and are higher than the rates for many other types of bonds. State and local governments issue "short-term municipals" of all types, traditional and new, to meet capital needs while waiting for long-term funding revenues. These bonds are issued to fund activities such as urban renewal and wastewater treatment.

Reference for Further Information: The Bond Market Association; Temel, Judy W.; *The Fundamentals of Municipal Bonds*, 5th ed., John Wiley & Sons, Inc.; 2001, available at http://www.amazon.com. Answers.com: http://www.answers.com/topic/municipal-bond.

Special Assessment Bonds

Description: Special assessment bonds are issued by local governments and/or special authorities and are secured by special taxes, charges, or fees. These bonds are sold to finance specific public infrastructure improvements that directly benefit the property owners in limited, identifiable areas. Assessments are levied on properties in the areas in direct relation to the benefits received from the projects. The assessments are based on property measurement systems related to the benefits such as street front-footage or square footage owned. The system for collecting assessments is usually tied to the collection of *ad valorem* property taxes, which are taxes based on the assessed value of property. Most special assessment bonds have maturities of 15 years or less. Examples of projects funded by special assessment bonds include the construction, maintenance, and/or repair of water and sewer lines, storm drains, sidewalks, and roadways, and public improvements including parks, bicycle paths, and landscaping.

Reference for Further Information: "Special Tax Bonds" and "Tax Increment Bonds" in this section of the Guidebook. Standard and Poor's Corporation, *Standard and Poor's Municipal Finance Criteria,* S & P Publications: 1993. The Bond Market Association; Temel, Judy W.; *The Fundamentals of Municipal Bonds*, 5th ed.; John Wiley & Sons, Inc.: 2001, "special assessment bonds" listed in index under "Bonds," available at http://www.amazon.com.

Special Tax Bonds

Description: Special tax bonds are backed by pledges of proceeds from specific tax sources. They are usually issued by local governments to finance specific types of facilities or environmental protection initiatives. For environmental purposes, particularly for the financing of parks and open space purchases, localities use special tax bonds financed out of local sales tax surcharges or property tax surcharges. Such surcharges may be approved for a limited time period or to collect a specified amount of money. Special tax bonds have long been issued by highway authorities, paid for out of highway taxes, to finance highways, roads and bridges. Special tax bonds combine some of the characteristics of revenue bonds, general obligation bonds, and special assessment bonds.

Reference for Further Information: "Revenue Bonds," "General Obligation Bonds," and "Special Assessment Bonds" in this section of the Guidebook. The Bond Market Association; Temel, Judy W.; *The Fundamentals of Municipal Bonds*, 5th ed.; John Wiley & Sons, Inc.; 2001, "special-tax" is listed in the index under "Bonds," available at http://www.amazon.com.

State Revolving Fund (SRF) Revenue Bonds

Description: State Revolving Fund (SRF) revenue bonds are issued to expand, or leverage, loan funding sources for local projects that meet the eligible project criteria under the Clean Water State Revolving Fund (CWSRF) and the Drinking Water State Revolving Fund (DWSRF). States use SRF dollars as security or as a source of revenue for the payment of principal and interest on bonds. Although SRF revenue bonds are issued at market rates, local borrowers receive loans at below market interest rates, subsidies provided in part by investments of the large bond debt reserve funds. SRF revenue bonds may be issued to provide for the required 20% state match to SRF federal capitalization grants. Many CWSRFs have received AAA ratings for their SRF revenue bonds. The bond leveraging approach has resulted in more loans being made, and more projects being funded, compared to the direct loan approach.

Reference for Further Information: See "U.S. Environmental Protection Agency: Clean Water State Revolving Fund" and "U.S. Environmental Protection Agency: Drinking Water State Revolving Fund" in Guidebook Section 2b.
"Fitch Rates Missouri SRF's $23.5MM Revenue Bonds 'AAA,'" Business Wire, Sept. 23, 2005, available at http://www.findarticles.com/p/articles/mi_m0EIN/is_2005_Sept_23/ai_n15627483.
Water Infrastructure Network Website:
http://win-water.org/legislativecenter/103101farrell.shtml. Oklahoma Water Resources Board
Website: http://www.owrb.state.ok.us/financing/news/fanews.php.

Tax Increment Bonds

Description: Tax increment bonds, which differ slightly from special assessment bonds, are tax-exempt bonds issued by local governments for special assessment or improvement districts where the benefit from the project being financed is specifically manifested through higher property values. Tax increment financing is used to generate revenue for bond repayment from the incremental change in property values caused by the financed improvement. After the creation of a special financing district by ordinance, two sets of tax records are maintained, one that reflects the property's value before the enhancement, and a second that reflects growing assessed values (and payments) after the enhancement and serves as the source of bond repayment. Tax increment financing bonds for revitalization projects may be backed by revenue pledges in addition to anticipated increases in property value, called "value capture," which makes them highly leveraged.

Reference for Further Information: "Tax Increment Financing" in Guidebook Section 8. "Special Assessment Bonds" in this section of the Guidebook. The Bond Market Association; Temel, Judy W.; *The Fundamentals of Municipal Bonds*, 5th ed.; John Wiley & Sons, Inc.; 2001, "tax-increment" listed under "Bonds" in index, available at http://www.amazon.com.

1. U.S. Department of Agriculture Business and Cooperative Programs: Economic Development Loans
2. U.S. Department of Agriculture Business and Cooperative Programs: Renewable Energy and Energy Efficiency Program
3. U.S. Department of Agriculture Rural Utilities Service: Water and Waste Disposal Systems Loans for Rural Communities
4. CoBank Loan Programs
5. Commercial Loans
6. U.S. Environmental Protection Agency: Clean Water State Revolving Fund
7. U.S. Environmental Protection Agency: Drinking Water State Revolving Fund
8. U.S. Environmental Protection Agency: State Revolving Fund Pre-Financing and Short-Term Loans
9. U.S. Environmental Protection Agency: Clean Water State Revolving Fund Private Beneficiary Loans
10. North American Development Bank
11. Private Investment
12. Environmental State Revolving Funds

U.S. Department of Agriculture Business and Cooperative Programs: Economic Development Loans

Description: These zero interest U.S. Department of Agriculture loans are used to promote sustainable rural economic development and job creation projects. Loans may be used to fund business expansions, startups, and incubator projects, community facilities, medical facilities, community infrastructure necessary for economic development and job creation purposes, and educational facilities and equipment. Electric and telephone utilities, and third parties applying through those utilities, are eligible for consideration for these loans. Examples of projects funded with these loans include the establishment or expansion of factories or businesses, medical facilities, and water and sewer industrial development parks. Most of the environmental projects funded involve water or wastewater systems. These loans could be used to help finance directly and leverage other capital for additional wastewater and drinking water utilities, and to fund non-point source improvements. The maximum loan amount is $740,000.

Reference for Further Information: U.S. Department of Agriculture Business and Cooperative Programs Website: http://www.rurdev.usda.gov/rbs/busp/redl.htm. These loans are also listed in the *Catalog of Federal Domestic Assistance (CFDA),* and on the Catalog's Website at http://12.46.245.173/cfda/cfda.html, search on program # 10.854.

U.S. Department of Agriculture Business and Cooperative Programs: Renewable Energy and Energy Efficiency Program

Description: The Farm Security and Rural Investment Act of 2002 established the Renewable Energy Systems and Energy Efficiency Improvements Program under Title IX, Section 9006. This program provides direct loans, loan guarantees, and grants to agricultural producers and rural small businesses to assist them with purchasing renewable energy systems and making energy efficiency improvements. The minimum amount of these loans is $5,000, less any project grant amounts, and the maximum loan amount is $10 million. Funds are awarded through this program for the installation of a wide range of wind, solar, biomass, geothermal, and conservation technologies. Guaranteed loans can be used for making capital improvements to existing renewable energy systems, but direct loans cannot.

Reference for Further Information: U.S. Department of Agriculture Website: http://www.rurdev.usda.gov/rbs/farmbill/. Applications should be submitted to the U.S. Department of Agriculture Rural Development state office in the state where the project is located. This program is listed on the Catalogue of Federal Domestic Assistance at http://12.46.245.173/cfda/cfda.html, search on program # 10.775.

U.S. Department of Agriculture Rural Utilities Service: Water and Waste Disposal Systems Loans for Rural Communities

Description: These U.S. Department of Agriculture loans provide assistance for meeting rural water and waste disposal needs in cities and towns with populations of 10,000 or less. The loans are intended for providing basic human amenities, alleviating health hazards, and promoting the orderly growth of the rural areas of the nation by meeting needs for new and improved rural water and waste disposal facilities. Funds may be used for the installation, improvement, or expansion of rural water facilities and the repair of distribution lines and well pumping facilities. In addition, funds may also be used for the installation, repair, improvement, or expansion of rural waste disposal facilities and the collection and treatment of sanitary, storm, and solid wastes. Eligible applicants include municipalities, counties, and other political subdivisions of a state, such as districts and authorities, associations, cooperatives, nonprofit corporations, and federally recognized tribes. No maximum loan amount has been established through statute.

Reference for Further Information: U.S. Department of Agriculture Website: http://www.usda.gov/rus/water/index.htm. These loans are also listed in the *Catalog of Federal Domestic Assistance (CFDA),* and on the Catalog's Website at http://12.46.245.173/cfda/cfda.html, search on program # 10.760.

CoBank Loan Programs

Description: CoBank is a private financial institution that offers a broad range of flexible loan programs, specially tailored financial services, and leasing services to agribusinesses, rural communications systems, Farm Credit Associations, and water, waste disposal, and energy systems. CoBank operates as a financial cooperative and is part of the Farm Credit System, a $140 billion national network of independently owned and operated credit and financial institutions providing services to agriculture and aquatic producers and rural homeowners. A broad range of competitively priced and flexible short-term, intermediate and long-term loan programs are offered through CoBank. Short-term loans, which usually mature within 12 to 18 months, are available through CoBank to finance current or seasonal assets, inventories, accounts receivable, commodities and other short-term needs. Intermediate and long-term loans are available for the construction of new facilities, the remodeling or expansion of existing facilities, land or equipment purchases, and the financing of other long-term assets and working capital. Short-term loans and intermediate and long-term loans are available at fixed and variable rates.

Reference for Further Information: CoBank Website: http://www.cobank.com/.
Farm Credit Services Website: http://www.farmcredit.com/.

Commercial Loans

Description: Most commercial banks and financial institutions in the United States have public finance departments that provide state and local governments with loans to finance a wide variety of capital projects and purchases. States and local governments tend to use commercial loans when lower-interest financing is unavailable and/or to fill short-term financing needs in anticipation of revenues from other sources (i.e., so-called bridge loans). Commercial loans are usually provided at set costs keyed within a range of market-based interest rates. They tend to have higher interest rates and less favorable payback terms as compared to government loans. Commercial lenders such as banks are very low-risk lenders and they usually seek to protect themselves and their loans by securing collateral in one or more of three ways: primary collateral in the form of assets (preferably liquid), secondary collateral such as guarantees, and cash flow. For governments, a portion of future revenues or taxes often represents the ultimate security for commercial loans. The application process for commercial loans tends to be much faster than for government loan programs. Commercial lenders usually have no set eligibility criteria and may have no predetermined limits on the total amounts of loan capital that they make available.

Reference for Further Information: Most commercial banks have public finance departments and/or environmental services departments that assist with inquiries on loan programs. Those that do not either handle inquiries from their general finance/loan operation departments or refer people to other banks.

U.S. Environmental Protection Agency: Clean Water State Revolving Fund

Description: Under Title VI of the 1987 Clean Water Act, states receive federal monies to capitalize Clean Water State Revolving Fund (CWSRF) loan programs. Through CWSRF programs, loans are made to communities to provide low cost financing for a wide range of different projects for the protection of water quality. Examples of activities funded with these loans include nonpoint source pollution control, watershed protection and restoration, estuary management, wetlands restoration, brownfields remediation, and improvements to municipal wastewater treatment infrastructure. Loans are made at low interest rates (0 percent to market rate) for terms of up to 20 years. In addition, states use CWSRF money to repurchase debt to get these loans to 30 years. States may set the criteria for determining which municipalities can access the loans each year. All 50 U.S. states and Puerto Rico operate CWSRFs.

Reference for Further Information: U.S. Environmental Protection Agency Office of Wastewater Management Website: http://www.epa.gov/owm/cwfinance/cwsrf/index.htm, phone 202-564-0752. See "Clean Water State Revolving Fund Private Beneficiary Loans" and "State Revolving Fund Pre-Financing and Short-Term Loans" in this section of the Guidebook and "State Revolving Fund Revenue Bonds" in Guidebook Section 2a.

U.S. Environmental Protection Agency:
Drinking Water State Revolving Fund

Description: The 1996 Safe Drinking Water Act (SDWA) Amendments authorize funding for the Drinking Water State Revolving Fund (DWSRF) program to provide public water systems with the financing needed to further public health under the SDWA. The DWSRF program provides states with capitalization grants that may be used for two major purposes. The first purpose is capitalization of state revolving loan funds to provide low cost loans, with 20-30 year terms, to public water systems. The second purpose is funding of "set-asides." Activities eligible for DWSRF loans include: projects to upgrade or replace drinking water treatment, transmission, distribution, and storage facilities, and consolidation of drinking water systems. Initiatives that are funded with DWSRF set-asides include: technical assistance to water systems, drinking water source (source water) protection, and water system operator certification.

Reference for Further Information: U.S. Environmental Protection Agency Office of Groundwater and Drinking Water Website: http://www.epa.gov/safewater/dwsrf/index.html, phone 202-564-3750. See "State Revolving Fund Pre-Financing and Short-Term Loans" in this section of the Guidebook and "State Revolving Fund Revenue Bonds" in Guidebook Section 2a.

U.S. Environmental Protection Agency:
State Revolving Fund Pre-Financing and Short-Term Loans

Description: Some Clean Water State Revolving Fund (CWSRF) and Drinking Water State Revolving Fund (DWSRF) loan programs make short-term loans for planning, design and initial construction in localities which may later receive long-term CWSRF and DWSRF loans. In addition, State Revolving Fund loans may be used to pre-finance other federal or state drinking water loans or grants. State Revolving Fund pre-financing loans have been used for Rural Utility Service wastewater loans, Housing and Urban Development wastewater grants, and state loans and grants. The State of New York's short-term financing program provides interest free financing, for terms of up to three years, to recipients developing projects eligible for long-term DWSRF financing. Also, the State of Texas offers short-term, variable rate CWSRF loans that can be converted to long-term, fixed rate CWSRF loans at any time prior to project completion.

Reference for Further Information: U.S. Environmental Protection Agency Office of Wastewater Management Website: http://www.epa.gov/owm/cwfinance/cwsrf/index.htm, phone 202-564-0752. New York Department of Health Website:
http://www.health.state.ny.us/nysdoh/water/final/2005/2005iup.htm.
Texas Water Development Board Website:
http://www.twdb.state.tx.us/assistance/financial/fin_infrastructure/cwsrffund.asp

U.S. Environmental Protection Agency:
Clean Water State Revolving Fund Private Beneficiary Loans

Description: The Clean Water State Revolving Fund (CWSRF) program, unlike the Drinking Water State Revolving Fund (DWSRF) program, is statutorily limited to awarding loans for publicly owned projects. However, occasionally CWSRF loans are made through a municipal lease arrangement that allows the private sector to use the funds, as defined under the federal tax code. Under this type of arrangement, a CWSRF makes loans to a publicly owned entity, state or municipal, which has leased a facility to an entity in the private sector. The public entity acts as a conduit for loan funds to the private beneficiary. The private beneficiary makes lease and/or loan payments to the public entity through an operating lease or service agreement. The funds used for CWSRF private beneficiary lending, called economic development loans, are derived only from SRF "retained earnings," comprised of direct loan interest repayments and investment earnings on recycled dollars, as opposed to federal capitalization grant dollars. Thus, the number of such loans is automatically capped by the amount of retained earnings annually.

Reference for Further Information: See "U.S. Environmental Protection Agency: Clean Water State Revolving Fund" in this section of the Guidebook. State Revolving Funds program management should consult their U.S. Environmental Protection Agency (EPA) Regional Offices before awarding private beneficiary loans. U.S. EPA Office of Wastewater Management Website: http://www.epa.gov/owm/cwfinance/cwsrf/index.htm, phone 202-564-0752.

North American Development Bank

Description: The North American Development Bank (NADBank), and its sister organization, the Border Environmental Cooperation Commission (BECC), were created in 1994 under the auspices of the North American Free Trade Agreement (NAFTA). NADBank and the BECC were created for the purpose of funding environmental infrastructure projects along the United States/Mexico border. NADBank is a bilaterally-funded, international organization, capitalized and governed equally by the United States and Mexico. The BECC reviews proposals for environmental projects in the region along the United States/Mexico border and certifies them for loan funding by the NADBank. The mission of NADBank is to serve as a binational partner and catalyst in communities along the United States/Mexico border in order to enhance the affordability, financing, long-term development, and effective operation of infrastructure that promotes a clean, healthy environment for the people of the region. NADBank can provide financial assistance to public and private entities involved in developing environmental infrastructure projects in areas near the border. Projects financed by the NADBank must address environmental issues within 100 kilometers of either side of the United States/Mexico border.

Reference for Further Information: North American Development Bank Website: http://www.nadb.org/. See "Border Environmental Cooperation Commission" in Guidebook Section 5. Border Environmental Cooperation Commission Website: http://www.cocef.org/.

Private Investment

Description: Private investment is defined here as loans and other financial assistance originating from sources other than commercial banks and/or finance companies. Sources of private investment can include, but are not limited to, insurance companies, pension funds, venture capital funds, individual venture capitalists, corporation partners, and general capital investors. Private investment funds billions of dollars worth of new business start-ups in the United States each year. The entrepreneurial ventures funded with this private investment include the environmental goods and services sector as well as other environmental protection related activities. The potential uses of private investment for supporting environmentally-related businesses and/or activities are only limited by the degree of profit associated with them. If it can be demonstrated that an idea or activity will make money, then private investment can be found to support it. The application process for private investment is typically much faster than for government loan programs. Private investors usually have no set eligibility criteria and may have no predetermined limits on the total amount of loan capital available. Private investors tend to demand a significantly higher rate of return on their money than other sources of capital.

Reference for Further Information: Funding information on venture capital funds is available in directories such as, *Who's Who in Venture Capital*, April 2005, Grey House publishing, available at: http://www.greyhouse.com/venture_ww.htm. Venture Capital Resource Directory: http://www.vfinance.com/home.asp?ToolPage=vencaentire.asp.

Environmental State Revolving Funds

Description: Environmental revolving funds are state run lending institutions that are often modeled after the Clean Water State Revolving Fund (CWSRF) and the Drinking Water State Revolving Fund (DWSRF) described in this section of the Guidebook. Many states throughout the U.S., including Michigan, Kentucky, Ohio, and Alaska, have environmental revolving funds awarding loans for a wide variety of environmental protection initiatives including water pollution prevention, wastewater treatment, and brownfields revitalization. Environmental revolving funds allow states to plan and target limited resources to their highest priority needs.

Reference for Further Information: Council of Infrastructure Financing Authorities Website: http://www.cifanet.org/, see the publications list on the Website. Small Business Environmental's directory of state financial assistance programs: http://www.smallbiz-enviroweb.org/funding/fundstat.html. Michigan Department of Environmental Quality Website: http://www.michigan.gov/deq/0,1607,7-135-3307_3515_4143---,00.html. Kentucky Infrastructure Authority Website: http://kia.ky.gov/loan/. Ohio Environmental Protection Agency Website: http://www.epa.state.oh.us/cleanohio.html. Alaska Department of Environmental Conservation Website: http://www.dec.state.ak.us/water/muniloan/index.htm

1. U.S. Department of Agriculture Forest Service: Cooperative Forestry Assistance Grants
2. U.S. Department of Agriculture Rural Development: Grants for Water and Wastewater Revolving Funds
3. U.S. Department of Agriculture Forest Service: Forest Stewardship Program Grants
4. U.S. Department of Agriculture Forest Service: Urban and Community Forestry Program
5. U.S. Department of Agriculture Business and Cooperative Programs: Rural Business Enterprise Grants
6. U.S. Department of Agriculture Rural Utilities Service: Solid Waste Management Grant Program
7. U.S. Department of Agriculture Rural Utilities Service: Technical Training and Assistance Grant Program
8. Appalachian Regional Commission Grants
9. U.S. Department of Commerce Economic Development Administration: Public Works and Economic Development Grants
10. U.S. Department of Commerce National Oceanic and Atmospheric Administration: Coastal Services Center Grants
11. U.S. Department of Commerce National Oceanic and Atmospheric Administration: Coastal Zone Management Act Administration Awards
12. U.S. Environmental Protection Agency: Environmental Education Grant Program
13. U.S. Environmental Protection Agency: Environmental Justice Small Grants Program
14. U.S. Environmental Protection Agency: Performance Partnership Grants
15. U.S. Environmental Protection Agency: Program Grants
16. U.S. Environmental Protection Agency: Section 319 Nonpoint Source Pollution Control Grants
17. U.S. Environmental Protection Agency: Superfund Technical Assistance Grants
18. U.S. Environmental Protection Agency: Leaking Underground Storage Tank Trust Fund Grants
19. U.S. Environmental Protection Agency: Underground Storage Tank Categorical Grants
20. U.S. Environmental Protection Agency: Wetlands Program Development Grants
21. U.S. Environmental Protection Agency National Center for Environmental Research: Science to Achieve Results (STAR)
22. U.S. Environmental Protection Agency: Drinking Water State Revolving Fund Loan Principal Forgiveness
23. U.S. Department of Homeland Security Federal Emergency Management Agency: Superfund Amendments and Reauthorization Act, Title III Grants
24. Foundation and Corporate Giving
25. U.S. Department of Housing and Urban Development: Community Development Block Grant Entitlement Communities Grants
26. U.S. Department of Housing and Urban Development: State Community Development Block Grants Program for Non-Entitlement Areas

27.   U.S. Department of Interior Fish and Wildlife Service: Standard Grants Program for Wetlands Protection

28.   U.S. Department of Interior Fish and Wildlife Service: North American Wetlands Conservation Act Small Grants Program

29.   U.S. Department of Transportation: Transportation Equity Act for the 21st Century

U. S. Department of Agriculture Forest Service: Cooperative Forestry Assistance Grants

Description: The U. S. Department of Agriculture Cooperative Forestry Assistance program provides formula grants and project grants to state forestry agencies to assist in the advancement of forest resources management with respect to non-federal forests and other rural lands. Among the program's objectives are encouragement of the production of timber, control of insects and diseases affecting trees and forests, control of rural fires, improvement and maintenance of fish and wildlife habitat, the planning and conducting of urban and community forestry programs, and efficient utilization of wood and wood residues, including the recycling of wood fiber. State agencies can provide these grant funds to state forestry or equivalent agencies for forest stewardship programs on private, state, local, and other nonfederal lands. These agencies can use the grant funds for a wide range of initiatives pursuant to the program's objectives listed above. The average size for an individual grant award is $1,000,000.

Reference for Further Information: U.S. Department of Agriculture (USDA) Website: http://www.fs.fed.us/r6/coop/programs/forms/grant_forms.htm, http://www.fs.fed.us/spf/coop/library/grantguide.pdf, and http://www.fs.fed.us/spf/coop/programs/eap/index.shtml. USDA phone 202-205-1657. These grants are listed on the Catalogue of Federal Domestic Assistance at http://12.46.245.173/cfda/cfda.html, search on program # 10.664.

U.S. Department of Agriculture Rural Development: Grants for Water and Wastewater Revolving Funds

Description: U.S. Department of Agriculture grants for water and wastewater revolving funds are awarded to private nonprofit organizations to capitalize revolving loan funds. The revolving loan funds are used by the nonprofits to make small, short-term loans to help fund expenses including pre-development costs for water and wastewater disposal projects. The loan funds are also used for short-term costs incurred for replacement equipment, small-scale extension of services, or other small capital projects that are not part of the regular operations and maintenance activities of existing water and wastewater systems. Grant applicants are required to have the legal capacity and authority to perform the obligations of the grant, and the necessary expertise and experience in making and servicing loans. Loan applicants must demonstrate that they are unable to finance the proposed projects with their own resources or through commercial credit at reasonable rates and terms. Facilities receiving the loans must primarily serve rural residents and rural businesses.

Reference for Further Information: U.S. Department of Agriculture Website: http://www.usda.gov/rus/water/index.htm. Search on program # 10.864 in the Catalog of Federal Domestic Assistance at http://12.46.245.173/cfda/cfda.html.

U.S. Department of Agriculture Forest Service: Forest Stewardship Program Grants

Description: The U.S. Department of Agriculture Forest Stewardship Program awards project grants for the purpose of managing nonfederal forest lands. Landowners of nonfederal lands, nonprofits, tribes, and other state, local, and private agencies acting through state foresters, equivalent state officials, or other official representatives are eligible to apply for these grants. The objectives of the program are to promote and enable the long-term active management of nonfederal forest lands to sustain the multiple values and uses that depend upon such lands. The program awards funding for the delivery of information and professional assistance to owners of nonfederal forest lands. Funds are also awarded for afforestation, reforestation, and active management of nonfederal forest lands, and improved forest seedling production and distribution. These grant awards range from $25,000 to $2,000,000, with an average award of $450,000.

Reference for Further Information: U.S. Department of Agriculture Forest Service Website: http://na.fs.fed.us/stewardship/index.shtm, phone: 202-205-6206. These grants are listed in the Catalogue of Federal Domestic Assistance at http://12.46.245.173/cfda/cfda.html, search on program # 10.678.

U.S. Department of Agriculture Forest Service: Urban and Community Forestry Program

Description: The U.S. Department of Agriculture Urban and Community Forestry Program awards project grants to assist state foresters, equivalent state agencies, interested members of the public, and private nonprofit organizations in implementing urban and community forestry programs. The program's objectives include planning, establishing, managing and protecting trees, forests, green spaces, and related resources in and adjacent to cities and towns to improve urban livability, link government, private, and grassroots organizations and resources, and engage people in citizen-based, grassroots volunteer efforts to assist in retaining and protecting their natural environment. All states, as well as the District of Columbia, Puerto Rico, the U.S. Virgin Islands, the Commonwealth of the Northern Mariana Islands, American Samoa, Guam, and other U.S. possessions and territories are eligible to apply for these grants. Each grantee is required to furnish a 50% match in the form of cash, services, or in-kind contributions.

Reference for Further Information: U.S. Department of Agriculture (USDA) Website: http://www.fs.fed.us/ucf/, USDA phone: 202-205-1657. These grants are listed on the Catalogue of Federal Domestic Assistance at http://12.46.245.173/cfda/cfda.html, search on program # 10.675.

U.S. Department of Agriculture Business and Cooperative Programs: Rural Business Enterprise Grants

Description: U.S. Department of Agriculture Rural Business Enterprise Grants provide funds for the financing or development of small or emerging businesses. These grants may be used for developing employment opportunities with private businesses and industries to improve the economy in areas and communities with populations of less than 50,000. Public bodies, private nonprofit corporations and federally recognized Indian tribes are potentially eligible to receive these grants for the purposes of assisting a business. The grants are not awarded directly to the business. Grant funds may be used for establishing revolving loan funds for various rural development related purposes, construction of buildings, construction of water supply and waste disposal facilities, and programs providing educational and job training instruction.

Reference for Further Information: U.S. Department of Agriculture Website: http://www.rurdev.usda.gov/rbs/busp/rbeg.htm. Call the Rural Business Service National Office Specialty Lenders Division at 202-720-1400. Search on program # 10.769 in the Catalogue of Federal Domestic Assistance at http://12.46.245.173/cfda/cfda.html.

U.S. Department of Agriculture Rural Utilities Service: Solid Waste Management Grant Program

Description: These U.S. Department of Agriculture grants provide assistance for meeting rural water and waste disposal needs in cities and towns with populations of 10,000 or less. These grants may be used for evaluating landfill conditions to determine threats to water resources. They may also be used for providing technical assistance and/or training to landfill operators, communities (to help them reduce the solid waste stream), and operators of landfills that are closed or will be closed in the near future. Eligible applicants include private nonprofit organizations that have been granted tax exempt status by the Internal Revenue Service and public bodies including local governmental-based multi-jurisdictional organizations. Applications for these grants are accepted from October 1 through December 31 of each calendar year. Recipients of these grants for fiscal year 2006 include the Association of Vermont Recyclers, the Center for Ecological Technology in Massachusetts, and the Central Arkansas Planning and Development District.

Reference for Further Information: U.S. Department of Agriculture Website: http://www.usda.gov/rus/water/SWMG.htm. These loans are also listed in the *Catalog of Federal Domestic Assistance (CFDA),* and on the Catalog's Website at http://12.46.245.173/cfda/cfda.html, search on program # 10.762.

U.S. Department of Agriculture Rural Utilities Service: Technical Training and Assistance Grant Program

Description: This U.S. Department of Agriculture (USDA) grant program provides assistance for meeting rural water and waste disposal needs in cities and towns with populations of 10,000 or less. The objectives of this federal grant program are to identify and evaluate solutions to water and waste disposal problems in rural areas, assist applicants in preparing applications for water and waste grants made through state offices, and improve operation and maintenance of existing water and waste disposal facilities in rural areas. Private nonprofit organizations that have tax exempt status with the Internal Revenue Service are eligible to apply for these grants. Funds from these grants may be used to: 1.) identify and evaluate solutions to water and waste problems of associations in rural areas, 2.) assist associations that have filed a pre-application with the USDA in the preparation of water and/or waste loan and/or grant applications, 3.) provide training to association personnel that will improve the management, operation, and maintenance of water and waste disposal facilities, and 4.) to pay expenses associated with providing technical assistance and/or training.

Reference for Further Information: U.S. Department of Agriculture Website: http://www.usda.gov/rus/water/tatg.htm. These loans are also listed in the *Catalog of Federal Domestic Assistance (CFDA),* and on the Catalog's Website at http://12.46.245.173/cfda/cfda.html, search on program # 10.761.

Appalachian Regional Commission Grants

Description: Appalachian Regional Commission (ARC) grants are awarded to states, public bodies, and private nonprofit organizations for projects that create opportunities for self-sustaining economic development and improved quality of life for the people of Appalachia. ARC grants are either administered by ARC or by a federal agency selected by the grantee. There are four different types of Appalachian Regional Commission grants that fund environmental protection related activities: regional development grants, area development grants, local development district assistance grants; and research, technical assistance, and demonstration project grants. Demonstration Project grants fund analysis of funding gaps in drinking water and wastewater projects in the region. Environmental protection related projects funded with the other three grant types listed above include water and wastewater treatment systems. Funds for these grants are appropriated annually to the ARC by Congress.

Reference for Further Information: Appalachian Regional Commission Website: http://www.arc.gov/index.do?nodeId=101. Search on program numbers 23.001 (regional development), 23.002 (area development), 23.009 (local development district assistance), and 23.011 (research, technical assistance, and demonstration projects) in the Catalog of Federal Domestic Assistance at http://12.46.245.173/cfda/cfda.html.

U.S. Department of Commerce Economic Development Administration: Public Works and Economic Development Grants

Description: Public Works and Economic Development Program grants are administered by the U.S. Department of Commerce Economic Development Administration. These grants fund projects that enhance regional competitiveness and promote long-term economic development in economically distressed areas. The purpose behind these grants is to help communities and regions revitalize, expand, and upgrade their physical infrastructure to attract new industry, encourage business expansion, diversify local economies, and generate or retain long-term private sector jobs and investment. U.S. states, political subdivisions of states, territories, cities, counties, Indian tribes, and consortiums of Indian tribes are eligible to apply for these grants. Grant eligible projects include construction and maintenance of water and sewer facilities, industrial park infrastructure improvements, construction of vocational facilities, redevelopment of brownfields, and eco-industrial (ecological) development. These grants cover 50%-100% of project costs depending on economic need as defined by the Department of Commerce.

Reference for Further Information: U.S. Department of Commerce Website: http://www.eda.gov/AboutEDA/Programs.xml. Search on program # 11.300 in the Catalogue of Federal Domestic Assistance at http://12.46.245.173/cfda/cfda.html.

U.S. Department of Commerce National Oceanic and Atmospheric Administration: Coastal Services Center Grants

Description: The National Oceanic and Atmospheric Administration (NOAA)'s Coastal Services Center provides funding each year for special projects undertaken by the coastal management community. Funding is awarded in the form of competitive grants and cooperative agreements. These grants fund projects aimed at addressing coastal management issues in ways that promote maintenance or improvement of natural resources while also allowing for economic growth. The Coastal Services Center, along with many other NOAA agencies, issues an annual Omnibus notice in the *Federal Register* that describes grant funding availability. The number, nature, and amount of these grants, and who is eligible, varies from year to year. In addition, the Coastal Services Center's announcements of available competitive grants and cooperative agreements, along with application deadlines, are available on the Website below.

Reference for Further Information: National Oceanic and Atmospheric Administration, Coastal Services Center Website: http://www.csc.noaa.gov/funding/.

U.S. Department of Commerce National Oceanic and Atmospheric Administration: Coastal Zone Management Act Administration Awards

Description: The National Coastal Zone Management (CZM) Program, authorized by the Coastal Zone Management Act of 1972, provides grants in the form of cooperative agreements to coastal states and territories for administering coastal management programs. The CZM Program is administered by the Department of Commerce, National Oceanic and Atmospheric Administration (NOAA). The main goal of the CZM program is to preserve, protect, develop, and where possible restore and enhance the resources of the nation's coastal zone. Only Coastal and Great Lakes states and territories with coastal management programs approved by the Secretary of Commerce are eligible to receive CZM program funding. Thirty-four coastal and Great Lakes states and territories are receiving funding. Illinois is the only potentially eligible state that is not receiving funding. The only entities that may apply for CZM Program grants are the designated lead agencies for the coastal zone management program in each state or territory. The average size of award to each state or territory is $2.25 million. Tribes and other third parties are not eligible to receive coastal zone management grants. However, any state or territory that chooses to can make its coastal zone management funds available to tribes.

Reference for Further Information: National Oceanic and Atmospheric Administration (NOAA) Website: http://coastalmanagement.noaa.gov/programs/coast_div.html. Call Elisabeth Morgan at NOAA at 301-713-3155, ext. 166. These grants are listed in the Catalogue of Federal Domestic Assistance at http://12.46.245.173/cfda/cfda.html, search on program # 11.419.

U.S. Environmental Protection Agency: Environmental Education Grant Program

Description: The National Environmental Education Act of 1990 authorizes the Environmental Education Grant Program. The program is administered by the U.S. Environmental Protection Agency (EPA). The goal of the program is to support environmental education projects that enhance the public's ability to make informed and responsible decisions that affect environmental quality. To be grant eligible, an environmental education project must be based on sound science and promote environmental stewardship. The project must enhance critical thinking, problem solving, and effective decision making skills and teach individuals to weigh various sides of an environmental issue to make informed and responsible decisions. Grantees must provide non-federal matching funds of at least 25% of the project being funded. Each year the U.S. EPA issues a Solicitation Notice, published in the *Federal Register* and available on the Website below, that includes all the necessary application forms and instructions. Colleges and universities, local and tribal education agencies, state education or environmental agencies, nonprofit organizations, and non-commercial educational broadcasting entities are eligible to apply for grants under this program.

Reference for Further Information: U.S. Environmental Protection Agency Website: http://www.epa.gov/enviroed/grants.html.

U.S. Environmental Protection Agency:
Environmental Justice Small Grants Program

Description: The U.S. Environmental Protection Agency (EPA) Office of Environmental Justice established the Environmental Justice Small Grants Program (EJSG) in 1994. The purpose of the EJSG is to support and empower communities that are working on solutions to local environmental and/or public health issues. The U.S. EPA defines Environmental Justice as the fair treatment and meaningful involvement of all people regardless of race, color, national origin, or income with respect to the development, implementation, and enforcement of environmental laws, regulations and policies. Organizations officially designated by the Internal Revenue Service as 501(c)(3) nonprofits, and nonprofits recognized by the state, territory, commonwealth, or tribe in which they are located, are eligible to apply for these grants. The applicant must demonstrate having worked directly with, or provided services to, the affected community. An "affected community," for the purposes of these grants, is a community that is disproportionately affected by environmental harms and risks and has a local environmental and/or public health issue that is identified in the grant proposal.

Reference for Further Information: U.S. Environmental Protection Agency Website: http://www.epa.gov/compliance/environmentaljustice/grants/ej-smgrants.html, phone 202-564-2515, or 800-962-6215.

U.S. Environmental Protection Agency:
Performance Partnership Grants

Description: Performance Partnership Grants (PPGs) are multi-program grants made to state and tribal agencies by the U.S. Environmental Protection Agency (EPA) from funds allocated and otherwise available for categorical grant programs. They provide states and tribes with the opportunity to combine funds from two or more categorical grants into one or more PPGs. PPGs are authorized by the 1996 Omnibus Consolidated Rescissions and Appropriations Act (PL 104-134). There are twenty environmental program grants eligible for inclusion as Performance Partnership Grants in fiscal year 2006. PPGs give states and tribes increased flexibility to address their highest environmental priorities, thus increasing equity and environmental incentives. They provide incentives for states and tribes to improve their environmental performance and draw links between program goals and outcomes. PPGs also cut administrative burdens/costs for recipients and EPA by reducing the numbers of grant applications, budgets, work plans and reports. States and tribes must develop environmental indicators and performance measures for PPGs.

Reference for Further Information: U.S. Environmental Protection Agency Website: http://www.epa.gov/ocir/nepps/pp_grants.htm.

U.S. Environmental Protection Agency: Program Grants

Description: U.S. Environmental Protection Agency (EPA) program grants are awarded for various purposes including state and local program research, demonstrations, development, and implementation. The dollar amount, application criteria, and requirements differ from grant to grant, depending on Congressional authorization, statutory goals, and internal EPA grant policies. Some grant programs are specifically authorized for a particular purpose, while other programs give significant discretion to the EPA office administering the grants. EPA grants fund programs for the protection of all environmental media, including air, water, and soil. A number of EPA grants are targeted to research and demonstration projects; other grants provide support for state and local program activities that coincide with federal environmental quality priorities. Applicants must compete for limited funds, and they must sign EPA grant agreements before receiving financial awards.

Reference for Further Information: U.S. Environmental Protection Agency Grant Awards Database: http://yosemite.epa.gov/oarm/igms_egf.nsf/HomePage?ReadForm. The *Catalog of Federal Domestic Assistance* at http://12.46.245.173/cfda/cfda.html contains detailed information on EPA program grants and other federal grant programs.

U.S. Environmental Protection Agency: Section 319 Nonpoint Source Pollution Control Grants

Description: Section 319(h) of the Clean Water Act authorizes formula grants for state, tribal, and territorial water quality agencies to fund projects and programs designed to help reduce nonpoint sources of water pollution within identified priority watersheds. State foresters, private landowners, and private nonprofit organizations may apply through state lead agencies for these Section 319 grants. Grantees must utilize these funds to implement U.S. Environmental Protection Agency approved nonpoint source pollution management programs. A 40 percent nonfederal match, in the form of supplies, equipment, and/or funding, must be provided by grantees. Regulatory and nonregulatory programs assessing the success of specific nonpoint source pollution control projects may be eligible for these grants. In addition, enforcement, technical assistance, training, technology transfer, demonstration projects, and monitoring for nonpoint source projects may be eligible. An example of a potentially eligible project is the implementation and evaluation of Best Management Practices (BMPs) for animal waste.

Reference for Further Information: U.S. Environmental Protection Agency (EPA) Office of Wetlands, Oceans, and Watersheds (OWOW) grants program phone: 202-566-1155, Website: http://www.epa.gov/owow/nps/319hfunds.html. A list of state NPS coordinators to contact for grant applications is available on OWOW's Website. "EPA Funds Available for Forestry Projects": http://www.fs.fed.us/spf/coop/library/EPA_Funds.pdf.

U.S. Environmental Protection Agency:
Superfund Technical Assistance Grants

Description: Superfund Technical Assistance Grants (TAGs) are authorized by the Superfund Amendments and Reauthorization Act of 1986. TAGs provide money for activities that help communities participate in decision making at eligible Superfund sites. Initial grant awards of up to $50,000 are available to qualified community groups so they can contract with independent technical advisors to interpret technical information about their Superfund site and help the community understand that information. The technical advisors interpret and explain technical reports, site conditions, and the U.S. Environmental Protection Agency's proposed cleanup proposals and decisions at Superfund sites. Additional funds beyond the $50,000 may be awarded for very large or complex sites. TAGs are available at Superfund Sites that are on, or proposed to be on, the U.S. Environmental Protection Agency's National Priorities List. They must be Superfund sites for which a response action has begun.

Reference for Further Information: U.S. Environmental Protection Agency (EPA) Website: http://www.epa.gov/superfund/tools/tag/, the site lists contact information for the TAG coordinator for each EPA regional office.

U.S. Environmental Protection Agency:
Leaking Underground Storage Tank Trust Fund Grants

Description: The Leaking Underground Storage Tank (LUST) Trust Fund Program, administered by the U.S. Environmental Protection Agency, provides project grants in the form of cooperative agreements to support state corrective action and enforcement programs that address releases from leaking underground storage tanks containing petroleum. Grant funds are used to provide resources for the oversight and cleanup of petroleum releases from leaking underground storage tanks where owners and operators are unknown or are unwilling or unable to take corrective actions themselves. States may also oversee responsible party cleanups. A ten percent state cost share is required. The program can be used not only to address the immediate problem of leaking underground petroleum storage tanks, but also to raise public awareness of the pollution threat to groundwater caused by leaks from these tanks.

Reference for Further Information: U.S. Environmental Protection Agency Website: http://www.epa.gov/OUST/ltffacts.htm and http://www.epa.gov/swerust1/20clenup.htm. Contact Lynn Depont at the U.S. EPA at 703-603-7148.

U.S. Environmental Protection Agency:
Underground Storage Tank Categorical Grants

Description: The U.S. Environmental Protection Agency (EPA) provides funding to states, tribes, and/or intertribal consortia through the Underground Storage Tanks (UST) categorical grants to encourage owners and operators to properly operate and maintain their USTs for the prevention and detection of releases caused by spills, overfills, and corrosion. EPA awards grants to states for new activities authorized by the Energy Policy Act of 2005. EPA recognizes that the size and diversity of the regulated community puts state authorities in the best position to regulate USTs. However, EPA will use funds for direct implementation of release detection and prevention programs on tribal lands where EPA is legally responsible for carrying out the UST program. State UST grant activities focus on developing state programs with sufficient authority and enforcement capabilities to operate in lieu of the federal UST program. UST state grant activities also focus on ensuring that owners and operators routinely and correctly monitor all regulated tanks and piping in accordance with UST regulations. There is a 25 percent matching requirement for state grants and no matching requirement for tribal or intertribal grants.

Reference for Further Information: U.S Environmental Protection Agency (EPA) Website: http://www.epa.gov/swerust1/overview.htm. Contact Lynn Depont at the EPA at 703-603-7148. These grants are listed on the Catalogue of Federal Domestic Assistance at http://12.46.245.173/cfda/cfda.html, search on program # 66.804.

U.S. Environmental Protection Agency:
Wetlands Program Development Grants

Description: Wetlands Program Development Grants (WPDGs) provide funding for projects that promote the coordination and acceleration of research, investigations, experiments, training, demonstrations, surveys, and studies related to the causes, effects, extent, prevention, reduction, and elimination of water pollution. U.S. Environmental Protection Agency (EPA) regional offices administer these grants. States, tribes, local governments, interstate associations, inter-tribal agencies, intertribal consortia, and national nonprofit, nongovernmental organizations are eligible to apply for WPDGs. WPDGs can be used to build and refine any element of a comprehensive wetlands protection program. Priority for funding is given to projects that address the three priorities identified by the U.S. EPA: 1.) developing a comprehensive monitoring and assessment program, 2.) improving the effectiveness of compensatory mitigation, and 3.) refining the protection of vulnerable wetlands and aquatic resources.

Reference for Further Information: U.S. Environmental Protection Agency (EPA) Office of Wetlands, Oceans and Watersheds (OWOW) Website: http://www.epa.gov/owow/wetlands/grantguidelines/, contact information for the EPA regional offices that administer these grants is available on the Website.

U.S. Environmental Protection Agency National Center for Environmental Research: Science to Achieve Results (STAR) Grants

Description: The Science to Achieve Results (STAR) program, administered by the U.S. Environmental Protection Agency's National Center for Environmental Research, funds research grants in numerous environmental science and engineering disciplines. The STAR program awards these grants to scientists and engineers at universities and nonprofit organizations, funding targeted research that complements EPA's intramural research program. Cutting edge science and proof of concept-type projects consistent with EPA's mission and vision for environmental protection are funded with these grants. Past research funded with STAR grants has covered a wide range of technology areas focusing on green chemistry and engineering. Green chemistry utilizes innovative chemical technologies that reduce or eliminate the use or generation of hazardous substances in the design, manufacture, and use of chemical products. Green Engineering is the design, commercialization, and use of economical processes and products while reducing the generation of pollution at the source and minimizing risk to human health and the environment. STAR grants average about $350,000 per year for three years.

Reference for Further Information: U.S. Environmental Protection Agency (EPA) Website, Environmental Technology Opportunities Portal: http://es.epa.gov/ncer/grants/.
U.S. EPA Website, Green Chemistry page: http://www.epa.gov/greenchemistry/.
U.S. EPA Website, Green Engineering page: http://www.epa.gov/oppt/greenengineering/.

U.S. Environmental Protection Agency: Drinking Water State Revolving Fund Loan Principal Forgiveness

Description: The 1996 Safe Drinking Water Act (SDWA) established the Drinking Water State Revolving Fund (DWSRF), administered by the U.S. Environmental Protection Agency and capitalized by federal capitalization grants awarded to states. The states use the grant dollars to award loans to localities for public water systems. The SDWA permits states to use up to 30% of their federal capitalization grants for loan principal forgiveness to communities that the states defined as disadvantaged. Principal forgiveness is, in effect, a grant. Principal forgiveness is available for publicly and privately owned public purpose drinking water projects.

Reference for Further Information: "U.S. Environmental Protection Agency: Drinking Water State Revolving Fund" in Guidebook Section 2b. U.S. Environmental Protection Agency (EPA) Website: http://www.epa.gov/safewater/dwsrf/nims/disadvrg.pdf. Also see the Federal Register Notice and the Fact Sheet under "DWSRF Program Rule" and the *Annual Report* and *Report to Congress* on the EPA Website at: http://www.epa.gov/safewater/dwsrf/index.html#guidance. Localities should consult their state DWSRF officials. Contact Peter Shanaghan at the EPA at 202-564-3848 with general questions.

U.S. Department of Homeland Security Federal Emergency Management Agency: Superfund Amendments and Reauthorization Act, Title III Grants

Description: The Superfund Amendments and Reauthorization Act (SARA), Title III provides grants to fund education and training in emergency planning, preparedness, mitigation, response, and recovery associated with hazardous chemicals. This funding is available to federally recognized Tribal Nations only. SARA, Title III requires Tribal Nations to establish State Emergency Response Commissions to administer hazardous materials planning and preparedness within tribal boundaries. Title III recognizes tribal responsibility as equal to states in terms of the protections of lives and property from chemical hazards. Individuals eligible for the training funded with Title III grants include public officials, fire and police personnel, medical personnel, first responders, and other tribal response and planning personnel. Training funded with these grants includes federal activities and conferences, state training programs, private sector training, university training centers, and other training sources. These grants may also be used to pay contractual services acquired for the purpose of training and educating the Tribal Nations.

Reference for Further Information: U.S. Federal Emergency Management Agency Website: http://www.fema.gov/government/grant/sara.shtm.

Foundation and Corporate Giving

Description: Foundation and corporate grants are a significant and growing source of funding for environmental protection projects. Most grants of this type fund well defined projects, with specified time frames, costs, and deliverables, that meet the immediate priorities of the funding source, and are not funded by governments. Foundation and corporate grant programs tend to favor the most innovative environmental projects. Grants are frequently awarded for environmental research, education, planning, and demonstration projects. Examples of organizations awarding grants for environmental protection projects include the Ben & Jerry's Foundation, Patagonia, Inc.; Ford Motor Company, and the Charles Stewart Mott Foundation.

Reference for Further Information: The Foundation Directory: http://fconline.fdncenter.org/. Foundation Center Website: http://foundationcenter.org/findfunders/. *Environmental Grantmaking Foundations,* 11th ed., available through http://www.environmentalgrants.com/. *National Directory of Corporate Giving*, 12th ed., available through the Foundation Center at http://foundationcenter.org/marketplace/. Ford Motor Company Website: http://www.ford.com/en/goodWorks/fundingAndGrants/conservationEnvironmental/default.htm. Charles Stewart Mott Foundation Website: http://www.mott.org/programs/environment.asp. Ben & Jerry's Foundation Website: http://www.benjerry.com/foundation/guidelines.html. Patagonia Website: http://www.patagonia.com/web/us/patagonia.go?assetid=2927.

U.S. Department of Housing and Urban Development:
Community Development Block Grant Entitlement Communities Grants

Description: The Community Development Block Grant Entitlement Communities Grants program seeks to develop viable urban communities by providing decent housing and a suitable living environment, and by expanding economic opportunities, principally for low and moderate income individuals. The program, administered by the U.S. Department of Housing and Urban Development (HUD), provides annual grants on a formula basis to cities and counties defined as entitlement communities. HUD defines entitlement communities as principal cities of Metropolitan Statistical Areas (MSAs), other metropolitan cities with populations of at least 50,000, and qualified urban counties with populations of at least 200,000 (excluding the population of entitlement cities located in such counties). Environmental protection related activities funded with these grants include the construction of water and wastewater treatment facilities, and activities related to energy conservation and renewable energy resources.

Reference for Further Information: The U.S. Department of Housing and Urban Development Website: http://www.hud.gov/offices/cpd/communitydevelopment/programs/entitlement/.
Search on program # 14.218 in the Catalogue of Federal Domestic Assistance at http://12.46.245.173/cfda/cfda.html.

U.S. Department of Housing and Urban Development:
State Community Development Block Grants Program for Non-Entitlement Areas

Description: The Community Development Block Grants (CDBG) Program for non-entitlement areas helps provide communities with decent housing, a suitable living environment, and expanded economic opportunities for low to moderate income citizens. States are grated authority by the U.S. Department of Housing and Urban Development (HUD) to administer these grants. The grants finance activities in non-entitlement areas, defined as cities with populations of less than 50,000 that are not principal cities of Metropolitan Statistical Areas (MSAs) and counties with populations of less than 200,000. Puerto Rico and all U.S. states except Hawaii receive funds to administer these grants to localities. HUD administers the program for non-entitled counties in the State of Hawaii because the state has permanently elected not to participate in the State CDBG Program. Environmental protection related initiatives funded with these grants include planning activities, acquisition of property for public purposes, and the construction of water and wastewater treatment facilities.

Reference for Further Information: U.S. Department of Housing and Urban Development (HUD) Website: http://www.hud.gov/offices/cpd/communitydevelopment/programs/stateadmin/.
Search on program # 14.228 in the Catalogue of Federal Domestic Assistance at http://12.46.245.173/cfda/cfda.html.

U.S. Department of Interior Fish & Wildlife Service: Standard Grants Program for Wetlands Protection

Description: The Standard Grants Program provides funding for public-private partnerships carrying out projects in the United States, Mexico, and Canada. To be eligible for funding through this program, projects must involve the long-term protection, restoration, and/or enhancement of wetlands and associated upland habitats for the benefit of waterfowl and other migratory birds that depend upon wetlands habitat. Projects carried out in Mexico may also include technical training, environmental education and outreach, organizational infrastructure development, and sustainable-use studies. The Migratory Bird Conservation Commission meets each year to approve the total amount of funding to be distributed to projects under the Standard Grants Program in the following fiscal year.

Reference for Further Information: U.S. Fish & Wildlife Service Website: http://www.fws.gov/birdhabitat/Grants/NAWCA/Standard/index.shtm, contact information for grant coordinators in the U.S., Canada, and Mexico are listed on the Website. This program is listed as the "North American Wetlands Conservation Fund" in the Catalogue of Domestic Assistance at http://12.46.245.173/cfda/cfda.html, search on program # 15.623. The program is also listed at: http://www.grants.gov/search/search.do?mode=VIEW&oppId=7818.

U.S. Department of Interior Fish & Wildlife Service: North American Wetlands Conservation Act Small Grants Program

Description: This Small Grants Program provides funding for public-private partnerships carrying out projects in the United States that further the goals of the North American Wetlands Conservation Act. To be eligible for funding through this program, projects must involve the long-term protection, restoration, and/or enhancement of wetlands and associated upland habitats for the benefit of waterfowl and other migratory birds that depend upon wetlands habitat. Grantees are required to come up with matching funds. The Migratory Bird Conservation Commission meets each year to approve the total amount of funding to be distributed to projects under the Small Grants Program in the following fiscal year.

Reference for Further Information: Contact Small Grants Coordinator Keith Morehouse at the U.S. Fish and Wildlife Service: phone 703-358-1888, e-mail: keith_morehouse@fws.gov.
U.S. Fish and Wildlife Service Website:
http://www.fws.gov/birdhabitat/Grants/NAWCA/Small/index.shtm.
This program is listed as the "North American Wetlands Conservation Fund" in the Catalogue of Domestic Assistance at http://12.46.245.173/cfda/cfda.html, search on program # 15.623. The program is also listed at: http://www.grants.gov/search/search.do?mode=VIEW&oppId=7818.

U.S. Department of Transportation: Transportation Equity Act for the 21ST Century

Description: The Transportation Equity Act for the 21st Century (TEA-21), signed into law in 1998, authorizes over $200 billion to improve the nation's transportation infrastructure, enhance economic growth, and protect the environment. State and local governments coordinate TEA-21 transportation project planning and funding processes. Through additions to both the Surface Transportation Program and the National Highway System, TEA-21 provides funding for environmental protection initiatives. TEA-21 provides this funding through grants and set-asides for a wide variety of water quality enhancement and transportation enhancement initiatives. Eligible water quality enhancement initiatives include wetlands mitigation banking, wetlands restoration, the mitigation of water pollution due to highway runoff, and the construction of pump out and dump station facilities in marinas. Eligible transportation enhancement activities include the construction of bicycle and pedestrian pathways, the acquisition of conservation or scenic easements, and rails-to-trails projects.

Reference for Further Information: U.S. Environmental Protection Agency Website: http://www.epa.gov/owow/tea/teafact.html. U.S. Department of Transportation Website: http://www.fhwa.dot.gov/tea21/implinks.htm and http://www.fhwa.dot.gov/tea21/sumover.htm.

Environmental protection needs and expectations are continuing to grow, while the resources available to meet those needs and expectations are increasingly constrained at all levels of government. Federal, state and local governments and the private sector are exploring the use of more efficient, effective, and innovative solutions to help address these major challenges. They are aggressively looking for and creating ways and opportunities to lower environmental costs, increase environmental investment, and build environmental capacity by creating partnerships with state and local governments and the private sector to fund environmental needs.

Federal, state, and local governments and the private sector are also looking for ways to enhance their credit so they can make their financial resources go further. Credit enhancement serves as an assurance to lenders and bondholders that credit is available, and that they will be repaid if the debtor government or private party should default or delay payment. Small businesses and local governments with poor credit ratings or no credit ratings may be able to gain access to capital markets and loan funds through credit enhancements, thus increasing the equity of access and allowing environmental projects to move forward.

This section presents and evaluates a number of the important mechanisms that these governments are testing and using to lower costs, enhance and build credit, increase investment, and build capacity through partnerships. It also looks at these mechanisms in terms of their contributions to financing environmental protection needs on a sustainable basis. The mechanisms reviewed in the section vary widely, ranging from specific analytical financial management tools to common-sense financial practices to broad, sweeping, innovative government programs and initiatives.

Some of the tools and initiatives discussed, such as refinancing and financial capability analysis, have been used for years. Others, such as cost-benefit analyses, cost-effectiveness analysis, and full-cost pricing are not new, but their use in the environmental arena may be new or growing. Still others, such as risk management and comparative risk ranking, have been used by one environmental media or by one level of government, and their use is now being incorporated in new areas or by new parties. One thing that these tools all have in common is that they represent approaches for lowering costs or enhancing credit, and for helping to address the long-term needs for financing environmental protection initiatives that are facing the nation.

1. Pooled Financing Programs
2. U.S. Small Business Administration: Surety Bond Guarantees
3. Bond Insurance
4. State Bond Banks
5. State Loan and Bond Guarantees
6. Performance Bonds
7. Letters of Credit
8. Lines of Credit
9. Senior and Subordinated Debt Structuring
10. State Revolving Fund (SRF) Interest Rate Subsidies
11. State Revolving Fund (SRF) Cross-Collateralization
12. Accelerated Depreciation
13. Activity-Based Costing
14. U.S. Department of Housing and Urban Development Community Development Block Grant Program: Section 108 Loan Guarantees
15. Amortization of Pollution Control Facilities
16. Appropriate Technology
17. Barter
18. Fiscal Impact Analysis
19. Full Cost Pricing
20. Expensing of Assets
21. Pay-As-You-Go
22. Refinancing Loans
23. Reforestation Tax Deduction and Amortization
24. Rehabilitation Tax Credits
25. Risk Management
26. Discounting
27. Value Engineering
28. Environmental Due Diligence
29. Financial Due Diligence
30. Benchmarking
31. Deduction of Agricultural Conservation Expenses
32. Comparative Risk Ranking
33. Cost-Effectiveness Analysis
34. Financial Capability Analysis
35. Cost-Benefit Analysis
36. Capital Planning and Budgeting
37. Employee Stock Ownership Plans
38. Life Cycle Assessment and Design

Pooled Financing Programs

Description: Pooling is the act of uniting, or an agreement to unite, an aggregation of properties belonging to different individuals, with a view to common liabilities or profits. Financing mechanisms such as loans and bonds are often pooled in arrangements called pooled financing programs or credit pools. Pooled financing programs can be used to lower interest rates, lower issuance costs, and increase flexibility for loans, bonds, and other financing mechanisms. Through credit pools, local government agencies and other organizations can take advantage of economies of scale by sharing the costs of issuing bonds or other debt instruments and securing lower interest rates. Regional credit pools are often used to finance local capital infrastructure projects such as water, sewer, and road improvements. The Florida Municipal Power Agency's Pooled Loan Project finances various electric, gas, water, sewer, and other utility related projects. The Association of Bay Area Governments (ABAG) in the San Francisco Bay Area manages 46 credit pools that have funded over 131 projects totaling more than $319 million. ABAG's credit pools have been used to fund a variety of projects, including construction and renovation of municipal and public safety buildings, and water, sewer, and drain projects.

Reference for Further Information: Florida Municipal Power Agency Website: http://fmpa.com/html/member_services/pooled_loan.html.
Association of Bay Area Governments Website:
http://www.abag.ca.gov/services/finance/pooling/summary.htm.

U.S. Small Business Administration: Surety Bond Guarantees

Description: The U.S. Small Business Administration (SBA) guarantees surety bonds for contracts up to $2 million, covering bid, performance, and payment bonds for small and emerging contractors who cannot obtain surety bonds through customary commercial channels. A surety bond guarantee is an agreement between a surety and the SBA. Surety bond guarantees provide that the SBA will assume a predetermined percentage of loss in the event that the contractor should breach the terms of the contract. The SBA's guarantees give sureties an incentive to provide bonding for eligible contractors, improving the contractors' chances of obtaining bonding. Contractors must qualify as small businesses based on the SBA's criteria and meet the surety's bonding qualifications to be potentially eligible for SBA surety bond guarantees. Professional agents or brokers specializing in providing surety bonds and the U.S. Department of Treasury, Financial Management Service, can provide information regarding specific sureties. Most large property and casualty insurance companies have surety departments.

Reference for Further Information: U.S. Small Business Administration (SBA) Website: http://www.sba.gov/services/financialassistance/basics/sbarole/surebond_geninfo.html. See "Surety Bonds" in Section 10b of this Guidebook and "Performance Bonds" in this Section of the Guidebook. U.S. Department of the Treasury Website: http://www.fms.treas.gov/c570/c570.html.

State Bond Banks

Description: State bond banks are public authorities created to help communities, especially smaller ones without substantial financial expertise or credit history, to access the reduced loan rates and other efficiencies of the tax-exempt bond market. By pooling smaller bond issues and providing state credit backing, state bond banks cut the cost of borrowing for communities. The Maine Municipal Bond Bank and the Virginia Resources Authority are examples of state bond banks. Bond banks are used to finance public facilities including wastewater and drinking water treatment systems and solid waste processing facilities. Bond banks provide three main economic advantages to localities: 1.) economies of scale in bond issuance, resulting from the elimination of duplication of fixed issuance costs, negotiated underwriting, administrative cost savings, and the use of specialized techniques to further reduce interest costs such as variable rates or zero coupon bonds, 2.) a pool of credit is generally perceived as more creditworthy than an individual credit because default risk is diversified, and 3.) improvements in credit quality via enhancement devices such as moral obligation pledges and revenue intercept mechanisms.

Reference for Further Information: Council of Development Finance Agencies Website: http://www.cdfa.net/cdfa/cdfaweb.nsf/pages/statebondbanksanderson.html. Maine Municipal Bond Bank Website: http://www.mainebondbank.com/. Virginia Resources Authority Website: http://www.virginiaresources.org/.

State Loan and Bond Guarantees

Description: Loan and bond guarantees are a form of credit assistance offered by states to recipients including counties, localities, and businesses. State guarantees are a very effective form of leveraging that can considerably reduce the costs of borrowing for bond issuers and loan recipients. Minnesota is an example of a state with a bond guarantee program. Minnesota's program reduces county borrowing costs on general obligation bonds by providing limited state guarantees of the payments on those bonds, thus allowing counties to receive higher bond ratings. Through the program, counties in Minnesota are provided with guarantees on general obligation bonds issued for the construction of solid waste facilities and other types of facilities specified by the state. An example of a state loan guarantee program is the Oregon Credit Enhancement Fund. This fund has guaranteed over 118 loans totaling over $27 million since 1994. The average credit enhancement loan awarded through this fund is for $230,000. Businesses and facilities meeting specific criteria are eligible for loan guarantees through the Credit Enhancement Fund. Any business that uses loan proceeds to clean up a brownfield site is eligible.

Reference for Further Information: Minnesota Department of Employment and Economic Development Website: http://www.deed.state.mn.us/programs/pfacountycred.htm.
State of Oregon Economic and Community Development Corporation Website: http://www.econ.state.or.us/cef.htm. See "General Obligation Bonds" in Section 2a of this Guidebook.

Performance Bonds

Description: Performance bonds are surety bonds issued by insurance companies on behalf of contractors, such as construction companies, to protect clients from the potential consequences of the contractors' failure to complete contracts in accord with plans and specifications. Performance bonds are frequently issued to secure the contactors' promises on the many public construction projects, such as wastewater treatment plant construction projects, that are performed by private sector firms in the United States. These bonds indicate that a financially responsible party, such as a commercial bank or insurance company termed the "surety," stands behind the contractor. Performance bonds permit the surety, upon contractor default, to either pay the bond penalty, or finance or hire a new contractor. Performance bonds are often required by the owners of the land to be developed or facility to be built. By furnishing these bonds, contractors may obtain credit, such as construction loans, at lower rates. These bonds limit surety liabilities to set amounts specified in bond agreements and contracts. Performance bonds can help environmental protection initiatives such as brownfields redevelopment projects which might otherwise be viewed as too risky or complex to move forward on a timely basis.

Reference for Further Information: Surety Information Office Website:
http://www.sio.org/html/why_bonds_reqd.html.
See Section 2a of this Guidebook for general information about bonds.

Letters of Credit

Description: Letters of credit (LCs) are documents that increase the basic security behind bonds and loans. LCs are generally issued by commercial banks and are used for two purposes: to enhance credit and enhance liquidity. A letter of credit represents a contract between the issuing bank and the bond trustee. In LCs issued for credit enhancement purposes, the issuing bank irrevocably agrees to provide funds to meet debt service payments in the event that the bond issuer or loan recipient is unable to do so and the LCs specify that funds will be used only for bond or loan repayment. In LCs designed for liquidity enhancement, the issuing bank provides liquidity enhancement by agreeing to advance any funds necessary to purchase bonds tendered by investors. LCs are used to increase market access for issuers who may have difficulty selling their bonds due to perceived weaknesses in their ability to meet their obligations. It is important that the bank issuing the LC has a sound financial history, a diverse loan portfolio, and adequate assets. Communities or companies that are ineligible for other types of credit enhancements could use LCs. LCs may be utilized to enhance the security behind bonds or loans used to finance environmental protection projects such as the construction of drinking water treatment plants or recycling facilities.

Reference for Further Information: See "Lines of Credit" in this section of the Guidebook. Temel, Judy W.; *The Fundamentals of Municipal Bonds*, 5th ed.; John Wiley & Sons, Inc.; 2001, pp. 190-191, available at http://www.amazon.com.

Lines of Credit

Description: A Line of Credit (LOC) is an arrangement between a bank or other financial institution and a customer that establishes a maximum loan balance that the financial institution will permit the borrower to maintain. LOCs are different from standard loans in that borrowers do not pay interest on the parts of their lines of credit that they do not use. Like letters of credit, lines of credit enhance the basic security behind debt instruments including bonds and loans. This is because the customer can access the LOC in times of financial hardship if needed to make payments on bonds and loans. The critical difference between letters of credit and lines of credit is that with lines of credit there are conditions under which the provider would not have advance funds under a draw. These conditions could potentially include default on the borrower's other debts, bankruptcy of the borrower, or a rating change, to name only a few. The analyst has to look at each individual line of credit to assess the degree of credit enhancement it provides. LOCs could be used by communities or companies that are ineligible for other types of credit enhancements. Lines of credit could potentially be utilized to enhance the security behind bonds used to finance environmental protection initiatives such as wastewater treatment plant construction.

Reference for Further Information: See "Letters of Credit" in this section of the Guidebook. Temel, Judy W.; *The Fundamentals of Municipal Bonds*, 5th ed.; John Wiley & Sons, Inc.; 2001, pp. 190-191. The Free Dictionary Website: http://financial-dictionary.thefreedictionary.com/Line+of+Credit+-+LOC.

Senior and Subordinated Debt Structuring

Description: Senior and subordinated debt structuring, also called subordinated debt, provides for two categories of debt for the loan recipient. The debt in the "senior" category is required to be repaid first should default or payment delays occur. The debt in the "subordinate" category is required to be repaid only after the senior debt, or lenders, are paid. This type of debt structuring is a credit enhancement for the senior debt or lender. Gladstone Capital is an example of a finance company that invests in senior subordinated loans. Subordinated debt structuring is used as a credit enhancement in State Revolving Fund (SRF) bond leveraged lending. For example, the New York State Environmental Facilities Corporation uses a senior subordinated debt structure for its Clean Water State Revolving Fund (CWSRF) and Drinking Water State Revolving Fund (DWSRF) revenue bonds. The Environmental Facilities Corporation uses the proceeds of offered bonds that are subordinated to senior bonds issued in its Master Financing Indenture (MFI) pooled financing program to finance or refinance water pollution control and drinking water projects.

Reference for Further Information: New York State Environmental Facilities Corporation Website: http://www.nysefc.org/docs/2006c.pdf. Gladstone Capital Website: http://www.gladstonecapital.com/profile_seniorsubdebt.htm. See "State Revolving Fund (SRF) Revenue Bonds" in Section 2a of this Guidebook. U.S. Environmental Protection Agency Website: http://www.epa.gov/owm/cwfinance/cwsrf/index.htm & http://www.epa.gov/safewater/dwsrf/index.html.

State Revolving Fund (SRF) Interest Rate Subsidies

Description: Interest rate subsidies on loans are a form of credit enhancement that is provided under the Clean Water State Revolving Fund (CWSRF) and Drinking Water State Revolving Fund (DWSRF) loan programs. Under federal statutes, states administering CWSRF and DWSRF loan programs are authorized to set specific loan terms, including interest rates, from zero percent to market rate, and repayment periods of up to twenty years. Interest rate subsidies on loans are, in effect, a credit enhancement for borrowers. This is because the subsidies reduce the cost of the loans, and thus increase the likelihood of loan applications being approved by increasing the perceived likelihood of repayment. These interest rate subsidies reduce the costs of environmental infrastructure for communities. The Texas Water Development Board is an example of an entity that offers interest rate subsidies on its CWSRF and DWSRF loans.

Reference for Further Information: See "U.S. Environmental Protection Agency: Clean Water State Revolving Fund," and "U.S. Environmental Protection Agency: Drinking Water State Revolving Fund" in Section 2a of this Guidebook. U.S. Environmental Protection Agency Website: http://www.epa.gov/owm/cwfinance/cwsrf/basics.htm & http://www.epa.gov/safewater/dwsrf/frequentquestions.html#3.
Texas Water Development Board Website: http://www.twdb.state.tx.us/assistance/financial/fin_infrastructure/cwsrffund.asp & http://www.twdb.state.tx.us/assistance/financial/fin_infrastructure/dwsrf.asp.

State Revolving Fund (SRF) Cross-Collateralization

Description: Cross-collateralization between the Clean Water State Revolving Fund (CWSRF) and the Drinking Water State Revolving Fund (DWSRF) was authorized by the Departments of Veteran's Affairs and Housing and Urban Development under the Independent Agencies Appropriations Act (Public Law 105-276) in 1999. The Act authorizes funds from one SRF program to be used to secure the other SRF program against default. This allows a DWSRF to benefit from existing CWSRF credit quality, diversification, and coverage levels, or vice versa. The U.S. Environmental Protection Agency (EPA) announced its policy on transfer and cross-collateralization of SRFs in the Federal Register on October 13, 2000. EPA emphasizes in the Federal Register notice that cross-collateralization can assist states in increasing the availability of SRF funds where they are most needed, enhancing bond ratings, and lowering borrowing costs without increasing risks. Through cross-collateralization, states may combine, or pool, assets of the CWSRF and the DWSRF programs and use them as security for bond issues. New Jersey is an example of a state that has received approval for a cross-collateralization structure and has issued bonds with the structure in effect.

Reference for Further Information: U.S. Environmental Protection Agency (EPA) Website: http://www.epa.gov/owm/cwfinance/cwsrf/innovations.htm, http://www.epa.gov/safewater/dwsrf/index.html and http://www.epa.gov/fedrgstr/EPA-WATER/2000/October/Day-13/w26353.htm. See "U.S. EPA Clean Water State Revolving Fund" and "U.S. EPA Drinking Water State Revolving Fund" in Guidebook Section 2b.

Activity-Based Costing

Description: Activity-based costing (ABC), a cost accounting methodology, is a method of allocating costs to products and services. It is generally used as a tool for planning. The use of ABC supports activity-based management that portrays an organization as a series of activities related to customer desires and costs. In ABC, functional costs, direct and indirect, are assigned to an organization's activities and those activities are traced to related products or services. Strategic management accounting based on the ABC methodology supports a long-term approach to decision making. In ABC, engineering process improvement is considered as a source of cost reduction. There are commercially available computer software packages for employing ABC on mainframes, networks, and personal computers. ABC is commonly used in cost accounting analyses, especially in the private sector. Information generated through the use of ABC gives visibility to how effectively resources are being used and how all relevant activities contribute to the costs of a product or service. Such information may help with decisions about whether to restructure or privatize environmental protection related activities.

Reference for Further Information: SAS Institute Website: http://www.sas.com/solutions/abm/. Infogoal.com: http://www.infogoal.com/category.php?n=19. Valuebased management.net: http://www.valuebasedmanagement.net/methods_abc.html. Accounting Software Research Website: http://www.asaresearch.com/articles/abc.htm.

U.S. Department of Housing and Urban Development Community Development Block Grant Program: Section 108 Loan Guarantees

Description: Section 108 is the loan guarantee provision of the U.S. Department of Housing and Urban Development (HUD) Community Development Block Grant (CDGB) program. Section 108 provides communities with financing for economic development, housing rehabilitation, public facilities, and large-scale physical development projects. All projects and activities funded through Section 108 are required to either principally benefit low and moderate income persons, aid in the elimination or prevention of slums and blight, or meet urgent community needs. Entitlement and nonentitlement communities, as defined by HUD, that are eligible for funds under the CDBG program, are potentially eligible for Section 108 loans. The maximum repayment time for Section 108 loans is twenty years. HUD has the ability to structure Section 108 principal amortizations to match the needs of projects and borrowers. Section 108 financing can be used for environmental protection purposes including construction of or modifications to public facilities such as wastewater treatment plants, improvements to streets and sidewalks that aid in storm water drainage, and redevelopment of abandoned sites that prevents the need to utilize open space for development projects.

Reference for Further Information: U.S. Department of Housing and Urban Development Website: http://www.hud.gov/offices/cpd/communitydevelopment/programs/108/. See the HUD Community Development Block Grant tools in Guidebook Section 2c.

Amortization of Pollution Control Facilities

Description: Amortization, similar to depreciation, is a method of recovering the capital costs of intangible assets over a fixed period of time by deducting them or writing them off. Under U.S. Internal Revenue Service (IRS) rules and U.S. federal law, the cost of a certified pollution control facility can be amortized over 60 months. A certified pollution control facility is defined by the IRS as a new identifiable treatment facility used in connection with a plant or other property in operation before 1976 to reduce the waste of resources and to prevent the creation of or reduce pollution and contamination to the water and atmosphere. The pollution control facility must be certified by state and federal certifying authorities. Amortization reduces taxable income for the affected tax years, thereby reducing the real cost of the pollution control facility by the amount that income taxes are reduced. For tax purposes, the distinction is not always made between amortization and depreciation. Still, amortization remains a viable financial accounting concept in its own right.

Reference for Further Information: Consult a tax practitioner.
Internal Revenue Service Website: http://www.irs.gov/publications/p535/ch08.html#d0e6593.
Cornell University Law School Website:
http://www.law.cornell.edu/uscode/html/uscode26/usc_sec_26_00000169----000-.html.
BNET Business Directory Website: http://dictionary.bnet.com/definition/Amortization.html.

Appropriate Technology

Description: The question of what technology is most appropriate must be made considering the environmental, cultural, and economic situations that the technology is intended for. Through the selection of appropriate technology, total life-cycle costs for that technology can be lowered substantially. Appropriate technology is selected based on realistic appraisals of input requirements and impacts on society and the environment. Basic types of technology that rely on locally available materials, geography, and resources may be most appropriate for certain areas and circumstances. For example, windmills are a less advanced form of technology than solar panels, but in very windy areas with little sunshine they would clearly be a more appropriate and cost effective choice for power generation. Choices among mixes of different technology characteristics work best in some situations. In some cases, organizations provide communities and groups with assistance in finding the appropriate technology. For example, the National Environmental Services Center (NESC) helps communities to determine the appropriate technology to meet their drinking water and wastewater treatment needs. Also, the National Center for Appropriate Technology serves economically disadvantaged people by providing them with information about and access to appropriate technologies.

Reference for Further Information:
National Environmental Services Center Website: http://www.nsfc.wvu.edu.
National Center for Appropriate Technology Website: http://www.ncat.org/about_history.html.

Barter

Description: Barter is a type of trade that does not use money or any other medium of exchange, but instead involves the trade of goods and/or services. Barter trade is also called reciprocal trade. The Internal Revenue Service (IRS) defines a barter exchange for tax purposes as "any person or organization with members or clients that contract with each other (or with the barter exchange) to jointly trade or barter property or services." Barter income is taxable under IRS rules, but the IRS does provide some exceptions to those tax rules, listed on the IRS website. There are a number of organizations involved in promoting barter. The International Reciprocal Trade Association is a nonprofit organization of companies dedicated to promoting just and equitable standards of reciprocal trade and raising the value of reciprocal trade to businesses and communities worldwide. The National Association of Trade Exchanges is an association of independent trade exchange owners who share their experiences and resources. Companies in the environmental goods and services industry, such as companies producing energy efficient appliances and equipment for reducing pollution, could use barter to preserve cash flow and transform surplus inventory into goods and services.

Reference for Further Information: International Reciprocal Trade Association Website: http://www.irta.com/Page.asp?Script=1. National Association of Trade Exchanges Website: www.nate.org/. Internal Revenue Service Website: http://www.irs.gov/businesses/small/article/0,,id=113437,00.html. Barter- Relevance and Relation to Money Website: http://www.ex.ac.uk/~RDavies/arian/barter.html.

Full Cost Pricing

Description: Full cost pricing is a method of factoring all current, past, and future costs; including operations, maintenance, and capital costs; into the prices for products and services. For example, public and private utilities such as water and wastewater treatment plants could utilize full cost pricing by setting user fees to recover all of the costs associated with maintaining facilities and providing services, including capital, operations, maintenance, debt service, and replacement. Full cost pricing could also be used in some jurisdictions to recover the costs of building and maintaining roads and highways. Rather than using full cost pricing, government entities often rely on tax dollars to partially or fully subsidize the costs of providing public services and facilities. For example, local governments may subsidize user fees for drinking water and wastewater utilities to assist low income households. Private businesses utilize the full cost pricing method much more frequently than government entities. Use of full cost pricing can help facilities and services to be financially self-sustaining. Full cost pricing systems that charge people more when they use more of a resource may encourage consumers to conserve valuable and limited natural resources such as water.

Reference for Further Information: U.S. Environmental Protection Agency Website: http://www.epa.gov/waterinfrastructure/pricing/About.htm & http://www.epa.gov/waterinfrastructure/fullcostpricing.html. Victoria Transport Policy Institute Website: http://www.vtpi.org/tdm/tdm29.htm.

Expensing of Assets

Description: Section 179 of the United States Internal Revenue Code allows businesses to immediately deduct the costs of certain types of property as expenses on their income taxes, rather than requiring the property to be capitalized and depreciated. This type of tax deduction is called expensing of assets. Under Section 179, businesses are allowed to elect current expense deductions in the year the qualifying property is placed in service, which gives them a more immediate tax benefits than does a depreciation deduction over a specified recovery period. Expensing is a widely and commonly used current-year income tax minimization strategy. The use of expensing increases current year cash profits by decreasing taxable income and consequent federal tax liability. Qualifying property is acquired for use in a trade or business and includes tangible personal property such as machinery and equipment. Assets that are used for environmental protection purposes, such as equipment for delivery of pollution remediation services, and equipment for generating renewable energy, including photovoltaic cells, solar hot water heaters, and windmills, can qualify for expensing of assets.

Reference for Further Information: Internal Revenue Service Website:
http://www.irs.gov/publications/p946/index.html. Consult a tax practitioner.
Cornell University Law School Website:
http://www.law.cornell.edu/uscode/uscode26/usc_sec_26_00000179----000-.html.

Pay-As-You-Go

Description: Pay-as-you-go is a system for funding public infrastructure that relies on tax and fee revenues, intergovernmental transfers, and/or trust fund balances rather than the issuance of debt. User fees and earmarked taxes tend to be the revenue foundations of pay-as-you-go systems. Specific percentages of general fund revenues are sometimes committed to pay-as-you go infrastructure projects for state and local governments. Reserve funds and trust funds are often utilized in pay-as-you-go systems to accrete sufficient cash amounts. The Highway Trust Fund, which provides financing for the national system of interstate and defense highways, is an example of that type of fund. Pay-as-you-go systems are often used to purchase assets such as communications equipment and transportation vehicles. In some jurisdictions, such as Buffalo, New York, pay-as-you-go systems are used to finance the capital costs of water and wastewater treatment systems. The pay-as-you-go approach is feasible for environmental protection projects that have sufficient political support to compete against other budget priorities.

Reference for Further Information: U.S. Department of Transportation Federal Highway Administration Website: http://www.fhwa.dot.gov/ and
http://www.fhwa.dot.gov/infrastructure/byrd.htm.
Northeast Midwest Institute Website: http://www.nemw.org/HWtrustfund.htm. City of Buffalo, Minnesota Website: http://www.ci.buffalo.mn.us/utilities/service/SACReport2004.pdf.
League of Women Voters of California Education Fund Website:
http://ca.lwv.org/lwvc/edfund/elections/2003/id/prop53.html.

Refinancing Loans

Description: Refinancing means applying for a secured loan to pay off an existing loan secured by the same assets. Through refinancing, monthly payments can be lowered on mortgages and other types of loans through a number of different methods. These methods include changing the loan to a lower interest rate, and extending the period of the loan, so as to extend the re-payment over a longer period of time. The most common consumer refinancing is done for home mortgages. Refinancing may be undertaken for the following reasons: 1.) to reduce interest rates on loans, 2.) to pay off other debts, 3.) to reduce periodic payment obligations (sometimes by taking out longer-term loans), 4.) to reduce risk by refinancing from a variable rate to a fixed rate loan, 5.) to liquidate some or all of the equity that has accumulated in real property during the tenure of ownership. Refinancing is done when the economic climate is one of lower interest rates compared to the time when the loan was originated. Most commercial lending institutions refinance loans. The refinancing need not be handled by the original lender. The money saved through refinancing of loans taken out for environmental protection purposes can be invested in additional environmental protection initiatives.

Reference for Further Information:
MortgageLoan Website: http://www.mortgageloan.com/refinance-mortgage.

Reforestation Tax Deduction and Amortization

Description: Internal Revenue Service (IRS) rules allow taxpayers in the United States to deduct a limited amount, up to $10,000 per year, of qualifying reforestation costs for each qualified timber property they own or lease. In addition, taxpayers can elect to amortize over 84 months any reforestation costs not deducted. There is no annual limit on the amount that taxpayers can elect to amortize. Amortization is a method of recovering (deducting) certain capital costs over a fixed period of time. Qualifying expenses for tax deductions and/or amortization are limited to costs which must otherwise be capitalized and included in the adjusted basis of the property, such as costs for site preparation, seeds or seedlings, labor, tools, and depreciation on equipment used in planting. To qualify for tax deductions and/or amortization, a timber property must be located in the United States, and it must consist of at least one acre planted with tree seedlings in the manner normally used in forestation or reforestation for commercial production of qualifying timber products. Tax deductions and amortization can make reforestation investments financially feasible for landowners.

Reference for Further Information: Consult a tax practitioner.
Internal Revenue Service Website: http://www.irs.gov/publications/p225/ch07.html & http://www.irs.gov/instructions/it/ar01.html. Private Landowner Network Website: http://www.privatelandownernetwork.org/yellowpages/resource.asp?id=6285.

Rehabilitation Tax Credits

Description: Federal tax credits to support private investment in the rehabilitation and reconstruction of historic buildings are authorized by the U.S. Internal Revenue Service (IRS) and U.S. Code, Title 26, Section 47. Generally, the percentage of expenditures that taxpayers are allowed to take as credits are 10% for buildings placed in service before 1936 and 20% for certified historic structures. For qualified rehabilitation costs paid or incurred after August 27, 2005, and before January 1, 2009, on buildings located in the Gulf Opportunity zone, the rehabilitation credit is increased from 10% to 13% for pre-1936 buildings, and is increased from 20% to 26% for certified historic structures. The term "certified historic structure" is defined as any building (and its structural components) which is listed in the National Register or is located in a registered historic district and is certified by the Secretary of the Interior as containing criteria which will substantially achieve the purpose of preserving and rehabilitating buildings of historic significance to the district. Renovation, restoration, and reconstruction activities qualify as rehabilitation under IRS rules. Enlargement and new construction do not qualify. These tax credits help to make rehabilitation and reconstruction of existing buildings more affordable, helping to prevent the need to utilize open space to build new buildings.

Reference for Further Information: Internal Revenue Service Website:
http://www.irs.gov/businesses/small/industries/article/0,,id=97599,00.html.
Cornell University Law School Website:
http://straylight.law.cornell.edu/uscode/html/uscode26/usc_sec_26_00000047----000-.html.

Risk Management

Description: Risk management involving the creation of pre-loss plans can be part of strategic planning that reduces costs and financial liabilities for organizations in the public and private sectors. Pre-loss plans are designed to minimize the adverse impacts of risks on the resources, earnings, cash flows, profitability, and credit ratings of organizations. These plans are most effective when they take into consideration public interest, public safety, and environmental protection. Financing and preparation techniques designed to help mitigate risks and prepare organizations for potential risks are included in pre-loss plans. In many cases, pre-loss plans involve risk transfer via the purchase of environmental insurance. Environmental insurance is a tool for managing a party's environmental liability by transferring some of the associated financial risk to another party under the limited provisions of the policy. Businesses in the environmental goods and services industry use risk management and pre-loss plans to help mitigate the risks associated with activities such as the cleanup of contaminated properties. The American Risk and Insurance Association is an organization devoted to the study and promotion of risk management and insurance.

Reference for Further Information: American Risk and Insurance Association Website:
http://www.aria.org/. See "Environmental Insurance" and "Environmental Risk Management in the Real Estate Industry" in Section 9 of this Guidebook.

Discounting

Description: In finance and economics, discounting is the process of finding the present value, also called the discounted value, that a given amount of cash will have at a future date. The present value of a cash flow is determined through the reduction of its value by an assigned discount rate, expressed as a percentage, for each unit of time between when the cash flow is to be valued and the present time. In most cases the discount rate is expressed as an annual rate. In financial accounting, discounting is the essence of most capital investment appraisal, comparing present cash flows with expected future cash flows. Discounting of an environmental protection measure is done by assigning a cash value and a discount rate to that environmental protection measure. For example, a cash value and a 10% annual discount rate could be used to estimate the present value that environmental economic benefits associated with salmon habitat restoration would have at a future date. The National Oceanic and Atmospheric Administration (NOAA) does discounting calculations of that nature to weigh the benefits and costs of coastal restoration projects and various environmental management programs. Discounting can provide a basis for making choices between investments in different environmental protection initiatives.

Reference for Further Information: National Oceanic and Atmospheric Administration (NOAA) Coastal Services Center Website: http://www.csc.noaa.gov/coastal/economics/discounting.htm.

Value Engineering

Description: Value engineering, also called value methodology, is defined in U.S. Public Law 104-106 as an analysis of the functions of a program, project, system, product, item of equipment, building, facility, service, or supply of an executive agency, performed by qualified agency or contractor personnel, directed at improving performance, reliability, quality, safety, and life cycle costs. Public Law 104-106 states that each executive agency shall establish and maintain cost-effective value engineering procedures and processes. The U.S. Environmental Protection Agency (EPA) defines value engineering as a specialized cost-control technique that uses a systematic and creative approach to identify and reduce unjustifiably high costs in a project without sacrificing the reliability or efficiency of the project. EPA is required to apply value engineering during Superfund Fund-lead remedial design and remedial action projects. Value engineering can be applied to various types of environmental protection projects in the public and private sectors, such as pollution prevention and cleanup projects. SAVE International is a society devoted to the advancement and promotion of value engineering.

Reference for Further Information: U.S. Environmental Protection Agency Website: http://www.epa.gov/superfund/cleanup/rdra.htm, see "value engineering" at the end of the list of topics. Text of Public Law 104-106: http://oecm.energy.gov/Portals/2/PL104_106.pdf. SAVE International Website: http://www.value-eng.org/.

Environmental Due Diligence

Description: Environmental due diligence is an investigation process carried out to ensure that a property is free of environmental contamination from current or past practices. One of the most important requirements for real estate transactions financed by institutional lenders is that environmental due diligence be done in a way that satisfies statutory requirements for conducting all appropriate inquiries as defined by the U.S. Environmental Protection Agency (EPA). The EPA published the All Appropriate Inquiries Final Rule (Final Rule) in the Federal Register on November 1, 2005. The Final Rule sets federal standards for the conduct of all appropriate inquiries. Starting on November 1, 2006, parties must comply with the requirements of the Final Rule, or with the standards of the Phase I Environmental Site Assessment Process, to satisfy the statutory requirements for conducting all appropriate inquiries. Adherence to the standards of the Final Rule or the Phase I Environmental Site Assessment Process is required for protecting property owners from liability under the Comprehensive Environmental Response, Compensation, and Liability Act (CERCLA).

Reference for Further Information: Bureau of National Affairs Website: http://www.bna.com/products/ens/eddg.htm. Environmental Data Resources Website: www.edrnet.com/reports/whitepapers/LendingtoSmall.pdf. Lawrence, Gary M., *Due Diligence in Business Transactions*, Law Journal Press, 1994; available at http://www.lawcatalog.com/product_detail.cfm?productID=1049&return=listview&setlist=0&. SCS Engineers Website: http://www.scsengineers.com/duedil.html.

Financial Due Diligence

Description: Financial due diligence is a series of tests that must be passed for a financing deal to qualify for an investment. The financial due diligence process is initiated when a business plan is sent to the potential investor, who applies preset criteria to screen out unacceptable deals. This screening includes a quantitative and qualitative analysis of how the business has performed financially to get a sense of earnings on a normalized basis. An identification and analysis of the assets and liabilities to be acquired is also included as part of the screening. In addition, an investigation is made into whether federal and state taxes have been filed appropriately by the seller. Financial due diligence is used by institutional investors and lenders considering commitment of funds to a venture.

Reference for Further Information: Lawrence, Gary M., *Due Diligence in Business Transactions*, Law Journal Press, 1994; available at http://www.lawcatalog.com/product_detail.cfm?productID=1049&return=listview&setlist=0&. Financial Evaluations & Examinations, Inc. Website: http://www.feeinc.com/burke.html. See "Business Plans" & "Venture Capital" in Guidebook Section 10a. Plante & Moran, PLLC Website: http://www.plantemoran.com/Services/Consulting/StrategyGlobalServices/Resources/Articles/Due+Diligence.htm.

Benchmarking

Description: Benchmarking is a management system in which organizations in the public and private sectors evaluate their processes and procedures in relation to best practices, develop plans on how to adopt the best practices, and establish measures of success called benchmarks. Most organizations find that benchmarking more than pays for itself. *Oregon Shines* is an example of a state run benchmarking system that is still in progress. It is a 20-year vision developed in 1989 to monitor the state's progress in emerging from an economic recession and improving public health and quality of life in the state. Connecticut, Florida, Maine, Minnesota, North Carolina, and Vermont have performance-based benchmarking systems similar to the Oregon model. An example of an environmental protection related benchmarking initiative is the survey that is being developed by the Book Industry Study Group (BISG) and the Green Press Initiative (GPI). The BISG and the GPI are working together, with support from industry sponsors, to produce a benchmarking survey that will establish a baseline for measuring the progress of the U.S. book industry in environmental protection related areas including paper recycling and climate impacts.

Reference for Further Information: Book Industry Study Group Website: http://www.bisg.org/publications/environmental_benchmarking.html.
Leichter, Howard M.; and Tryens, Jeffrey, *Achieving Better Health Outcomes: the Oregon Benchmark Experience*, Milbank Memorial Fund, New York, 2002, available at: http://www.milbank.org/reports/OregonProgres/020909Oregon.html.

Deduction of Agricultural Conservation Expenses

Description: U.S. Internal Revenue Service rules allow landowners and their tenants to deduct qualifying soil and water conservation expenses and erosion control expenses from their gross income for federal income tax purposes. These expenses must be incurred to protect land that the landowners or tenants are using or have used in the past for farming. The tax deduction cannot amount to more than 25% of the taxpayer's gross income from farming. Expenses can be deducted only if they are consistent with a plan approved by the U.S. Department of Agriculture's Natural Resources Conservation Service, or a soil conservation plan of a comparable state agency. Expenses for leveling, conditioning, grading, terracing, contour furrowing, restoration of soil fertility, and related treatment or movement of land are eligible. In addition, expenses for the construction, control, and protection of diversion channels, drainage ditches, irrigation ditches, earthen dams, watercourses, outlets, and ponds are eligible. Expenses for the planting of windbreaks and eradication of brush may also be deducted. To get the full deduction to which they are entitled, taxpayers must maintain records that clearly distinguish between their customary farm business expenses and their soil and water conservation expenses.

Reference for Further Information: Consult a tax practitioner.
Internal Revenue Service Website: http://www.irs.gov/publications/p225/ch05.html & http://www.irs.gov/businesses/small/industries/article/0,,id=99004,00.html. U.S. Department of Agriculture Natural Resources Conservation Service Website: http://www.nrcs.usda.gov/.

Comparative Risk Ranking

Description: Comparative risk ranking, also called comparative risk assessment and comparative risk analysis, is a procedure for prioritizing problems based on their threat to public health and the environment. The process of comparative risk ranking involves a high level of citizen input. Comparative risk ranking helps communities and organizations to allocate limited resources to the most serious environmental and public health problems first. For example, the Lower Columbia River Estuary Partnership integrated a comparative risk assessment into its management plan development to help assess potential actions. The Partnership's comparative risk assessment asked citizens and technical experts to rank a list of problems based on their threat to public health, ecological health, and quality of life. Another example of comparative risk ranking is the Ohio Comparative Risk Project. This was a citizen-based planning project that evaluated problems in Ohio based on scientific evidence and public values. The information collected in the study was then used to develop an environmental priority list and strategies for policymakers and citizens to use in reducing risk, made public in 2001.

Reference for Further Information: Lower Columbia River Estuary Partnership Website: http://www.lcrep.org/lib_risk_ranking.htm.
Scorecard Website: http://www.scorecard.org/comp-risk/def/comprisk_explanation.html.
U.S. Environmental Protection Agency Website: http://www.epa.state.oh.us/oeef/ohio_comparative_risk_project.html
Franklin Pierce Law Center Website: http://www.piercelaw.edu/risk/vol6/fall/kadvany.htm.

Cost-Effectiveness Analysis

Description: Cost-effectiveness analysis (CEA) is a form of economic analysis that compares the relative expenditures, called costs, to the outcomes, called effects, of two or more courses of action. For example, CEA may be done to compare the costs and effects of two different proposed regulations for improving air quality. CEA can also be used to measure the ratio of the costs of a single intervention, such as improvements to drinking water quality, to a measure of the intervention's effects, such as statistics on public health benefits in the community drinking the water. Cost-effectiveness analysis is done by federal departments and agencies, including the U.S. Environmental Protection Agency, when performing Regulatory Impact Analyses. Many state agencies and private businesses also perform CEA.

Reference for Further Information: U.S. Environmental Protection Agency (EPA), "Guidelines for Preparing Economic Analysis," 2000, EPA # 240-R-00-003, available at
http://yosemite1.epa.gov/ee/epa/eerm.nsf/vwSER/DEC917DAEB820A25852569C40078105B?OpenDocument.
AEI-Brookings Joint Center Website: http://www.aeibrookings.org/publications/index.php?tab=topics&topicid=56.
Environmental Damage Valuation and Cost Benefit Links:
http://www.costbenefitanalysis.org/tenbestedvcbnlinks.htm.
Environmental Valuation & Cost Benefit News: http://www.envirovaluation.org/.

Financial Capability Analysis

Description: Financial capability analysis, also called financial capability assessment, is a method used by public and private entities to determine whether they have the ability to pay for capital investments and equipment, and to assess the economic impacts of proposed projects on communities. The U.S. Environmental Protection Agency recommends that communities carry out financial capability analysis, using indicators including bond ratings and unemployment rates, to determine the affordability of infrastructure projects. Computer software, such as Boise State University Environmental Finance Center's Plan2Fund and CAPFinance software programs, is sometimes used for financial capability analysis.

Reference for Further Information: Financial Capability Guidebook, March, 1984, EPA #832B84104, available at http://yosemite.epa.gov/water/owrccatalog.nsf/, search the title index.
U.S. Environmental Protection Agency Website: http://www.epa.gov/npdes/pubs/csofc.pdf.
See "Boise State University Environmental Finance Center: Plan2Fund" and "Boise State University Environmental Finance Center: CAP Finance" in Section 5 of this Guidebook.
Boise State University Environmental Finance Center Website:
http://sspa.boisestate.edu/efc/Tools_Services/Plan2Fund/plan2fund.htm &
http://sspa.boisestate.edu/efc/Tools_Services/CAPFinance.htm.

Cost-Benefit Analysis

Description: Cost-benefit analysis is a term for a conceptual framework encompassing a variety of techniques for quantifying and comparing the incremental and total costs, risks, and benefits of legislation, regulations, and other actions. The use of cost-benefit analysis is intended to help produce the best decisions by comparing the economic efficiency of various proposed policies and approaches. Cost-benefit analysis has been used for many years by federal, state, and local governments, and by the private sector, as a tool to aid in important decision making on a wide variety of topics, including environmental protection related matters. The United States government has performed cost-benefit analysis as part of its economic analysis of regulatory actions for many years, following a number of statutory and executive order requirements including Executive Order 12866, titled *Regulatory Planning and Review*. Executive Order 12866 requires analysis of the benefits and costs of all significant regulatory actions carried out by the U.S. government and its agencies and departments, including the Environmental Protection Agency and other U.S. government entities working to protect the environment.

Reference for Further Information: U.S. Environmental Protection Agency Website:
http://yosemite.epa.gov/ee/epa/eed.nsf/webpages/Guidelines.html.
See the January 11, 1996 Office of Management and Budget policy memorandum, *Economic Analysis of Federal Regulations Under Executive Order* 12866, available at
http://www.whitehouse.gov/omb/inforeg/riaguide.html.
Mind Tools Website: http://www.mindtools.com/pages/article/newTED_08.htm.

Capital Planning and Budgeting

Description: Capital planning is a set of techniques for considering the long-term needs for capital facilities and funding options for meeting those needs. The goal of capital planning is to make the best use of available funds to achieve strategic goals and objectives. Capital budgeting, which is done alongside capital planning, is the total process of generating, evaluating, selecting, and following up on capital expenditures. Capital planning and budgeting are done in the public and private sectors. The Commission to Study Capital Budgeting released a report in 1999 with a series of recommendations to improve the federal budget process through prioritizing, making timely decisions, reporting on those decisions, and evaluating the decisions to help with future decision making. The commission's recommendations could help organizations in the public and private sectors to carry out their capital planning and budgeting more effectively and use their money wisely for environmental protection initiatives.

Reference for Further Information: The President's Commission to Study Capital Budgeting Website: http://clinton3.nara.gov/pcscb/index.html.
University of Wisconsin System Website: http://www.uwsa.edu/capbud/.
U.S. Department of the Interior Website: http://www.doi.gov/ocio/cp/index.html.
FEA DRM Registry Website:
http://colab.cim3.net/file/work/drm/schema/examples/IRM_COI_Demo/818.htm.

Employee Stock Ownership Plans

Description: An employee stock ownership plan (ESOP) is a type of defined contribution pension plan used in the United States that buys and holds company stock, investing primarily in the stock of the employer firm. ESOPs are required to adhere to the standards of the Employee Retirement Income Security Act of 1974 (ERISA), which is a U.S. federal law that sets minimum standards for most voluntarily established pension and health plans in private industry. ESOPs can be funded through tax deductible corporate contributions. Sellers to an ESOP in a closely held company can defer taxation on the proceeds by reinvesting in other securities. Employees do not pay taxes on the contributions to ESOPs until they receive a distribution from the plan when they leave the company. There are over 11,500 ESOPs in the U.S. covering 11 million employees, almost all in closely held companies. Employee owned corporations, which are defined as corporations owned in whole or in part by their employees, are often created through ESOPs. Companies in the Environmental Goods and Services industry, such as wind power generating companies and companies producing pollution abatement equipment, can use ESOPs to increase their profits and save on taxes.

Reference for Further Information: The National Center for Employee Ownership Website: http://www.nceo.org/. For tax rules, see the Internal Revenue Service Website: www.irs.ustreas.gov/. About.com: http://retireplan.about.com/cs/retirement/a/aa_plan_a6.htm.
U.S. Department of Labor Website: http://www.dol.gov/dol/topic/health-plans/erisa.htm.

Life Cycle Assessment and Design

Description: Life cycle assessment is defined by the U.S. Environmental Protection Agency as a "cradle to grave" approach for assessing industrial systems and products. "Cradle-to-grave" begins with the gathering of raw materials from the earth to create the system or product and ends at the point when all materials are returned to the earth. Life cycle assessment evaluates all stages of a product's life cycle from the perspective that they are interdependent, meaning that one operation leads to the next. By examining the impacts of a product or system throughout its life cycle, life cycle assessment provides a comprehensive view of the environmental impacts of the product or system and a more accurate picture of the true environmental tradeoffs in product and system selection. Life cycle assessments can help decision makers to select the products or processes that have the least impacts on the environment. Through the use of information collected in life cycle assessments, products and systems can be designed to reduce their environmental impacts throughout their life cycles. For example, measures can be taken to reduce the pollution created in the manufacture of products, the energy used through the manufacture and use of products, and the waste generated when the products are disposed of.

Reference for Further Information: U.S. Environmental Protection Agency Website: http://www.epa.gov/ORD/NRMRL/lcaccess/. Scientific Applications International Corporation (SAIC), Life Cycle Assessment: Principles and Practice, EPA/600/R-06/060, May 2006, available at Http://www.epa.gov/ORD/NRMRL/lcaccess/pdfs/600r06060.pdf.

Public-private partnerships are receiving increasing attention in the United States and internationally as an innovative way of financing a wide range of different environmental protection initiatives. The use of public-private partnerships to finance environmental protection programs and projects, and related infrastructure such as wastewater treatment plants, is examined in this section of the Guidebook. This section is divided into two subsections: Section 4A, "Tools for Building Public-Private Partnerships," and Section 4B, "Public-Private Partnership Case Studies." The tools describe how different types of public-private partnerships work. The case studies are demonstrations of how public-private partnerships can be used for various environmental protection purposes.

Some public-private partnerships presented in the tools and case studies fall into specific categories; such as contract services, turnkey, and privatization; while others do not. However, each public-private partnership has many unique characteristics that are closely examined. Many of the partnerships discussed are contractual relationships between government entities and private companies. Each public-private partnership covered in this section either helps to provide environmental protection related services or could potentially be used to help provide such services. Examples of environmental protection related services provided through public-private partnerships covered in the tools and case studies include wastewater treatment, delivery of clean drinking water, installation of solar electric systems, and purchases of land for conservation purposes.

In some cases, it is possible to capitalize on specific private sector resources through the use of public-private partnerships. The availability of those resources depends upon the nature of the partnership arrangements, the resources available to the private partners, the circumstances in the locations where they are set up, and other factors. Access to sophisticated technologies and specialized expertise often allows the private sector to provide specific types of services that the public sector may be unable to provide. In addition, private financing can reduce the burden on public debt capacity. Private sector procurement and construction methods sometimes save time and provide significant cost savings. Through public-private partnerships involving ownership transfers from government entities to private companies, responsibilities for financial risk can be transferred from the government entity to the private company.

There are some limitations involved with the use of public-private partnerships that must be considered. Local governments may not always have the legal authority to enter into contracts with private parties. A major concern of governments considering becoming part of public-private partnerships is the potential loss of oversight opportunities. When government officials cease to be involved with the day-to-day operations of a facility, they may have to give up opportunities to monitor things such as compliance with environmental standards and permits. In addition, public employees and unions may oppose the use of public-private partnerships due to concerns about the loss of jobs. Finally, tax-exempt and/or other low-cost financing that is available for federal and state government run projects may not be available for public-private partnerships.

The appropriateness of a particular type of public-private partnership for a given environmental protection initiative and location depends upon many factors such as the type of environmental media being protected, availability of public funding for the partnership, demographics, and the tax code. The tools and case studies included in this section are geared towards readers seeking to assess the benefits and limitations of different types of public-private partnerships. They contain important information intended to help the reader make decisions about how to use, and whether or not to use, public-private partnerships for specific environmental finance purposes. When structured and used appropriately for specific locations and circumstances, public-private partnerships make many important environmental protection initiatives financially possible.

1. Privatization
2. Asset Sales Under Executive Order 12803
3. Long-Term Lease Under Executive Order 12803
4. Tax-Exempt Lease
5. Lease/Purchase
6. Sale/Leaseback
7. Lease/Develop/Operate or Build/Develop/Operate
8. Build-Operate-Transfer
9. Contract Services: Operations and Maintenance
10. Contract Services: Operations, Maintenance, and Management
11. Turnkey
12. Developer Financing
13. Merchant Facilities

Privatization

Description: The simplest definition of privatization is the transfer of ownership of assets from a public sector organization, usually a state or local government, to businesses in the private sector. However, globally and within the United States, privatization has come to mean much more. The National Council for Public-Private Partnerships Federal Privatization Task Force defines privatization as a process of wide-ranging economic change that includes public-private partnerships, joint ventures, and outsourcing. The Task Force maintains that the change in ownership or control of the assets, or the prospect of it, is the catalyst for privatization, and that internal federal agency reorganization, absent the transfer of ownership, control, or responsibility to a private party, is not. Executive Order 12803, issued in 1992, defines privatization as "the disposition or transfer of an infrastructure asset, such as by sale or by long-term lease, from a state or local government to a private party." Many U.S. cities, such as the City of Hawthorne, California, have privatized their public wastewater and drinking water systems.

Reference for Further Information: The National Council for Public-Private Partnerships Website: http://www.ncppp.org/howpart/fedpriv.shtml#defpriv.
Executive Order 12803 as of April 30, 1992, available at
http://www.theantechamber.net/UsHistDoc/Exord12803/Exord12803Page1.html.
See "Asset Sales Under Executive Order 12803" in this section of the Guidebook.
ITT Industries Website: http://www.itt.com/waterbook/where.asp.

Asset Sales Under Executive Order 12803

Description: Executive Order 12803, issued in 1992, directs all United States federal departments and agencies to approve state and local governments' requests to privatize infrastructure assets financed in whole or part by the federal government to the extent permitted by law and consistent with originally authorized purposes. Privatization is defined in the Executive Order as "the disposition or transfer of an infrastructure asset, such as by sale or by long-term lease, from a state or local government to a private party." Executive Order 12803 clarifies the terms under which the federal government must be repaid for its investment upon the sale of a federally funded asset to the private sector. In effect, the Executive Order abolishes the requirement to repay the federal investment in full, greatly reducing the potential sales price. An example of a contract approved under Executive Order 12803 is Cranston, Rhode Island's lease of its entire wastewater treatment, collection, and pumping system to a private contractor for 25 years starting in 1997. This contract, the first in the United States where the contractor has full financial responsibility for a wastewater system, is expected to save the city $74 million.

Reference for Further Information: Executive Order 12803 as of April 30, 1992, available at
http://www.theantechamber.net/UsHistDoc/Exord12803/Exord12803Page1.html.
U.S. EPA Website: http://www.epa.gov/owm/cwfinance/privatization.htm.
ITT Industries Website: http://www.itt.com/waterbook/where.asp and
http://www.itt.com/waterbook/privatization.asp.
HDR Website: http://www.hdr-engineering.com/13/38/1/default.aspx?projectID=608.

Long-Term Lease Under Executive Order 12803

Description: Long-term leases are a form of privatization. Leases of ten years or more are considered long-term by investors. Executive Order 12803, issued in 1992, directs all United States federal departments and agencies to approve state and local governments' requests to privatize infrastructure assets financed in whole or part by the federal government to the extent permitted by law and consistent with originally authorized purposes. Privatization is defined in the Executive Order as "the disposition or transfer of an infrastructure asset, such as by sale or by long-term lease, from a state or local government to a private party." An example of a long-term lease approved under the authority of Executive Order 12803 is the City of Hawthorne, California's awarding of a 15 year lease to the California Water Service Company (Cal Water) for the management of its municipal water system starting in 1996.

Reference for Further Information: See "Asset Sales Under Executive Order 12803" and "Privatization" in this section of the Guidebook. Executive Order 12803 as of April 30, 1992, available at http://www.theantechamber.net/UsHistDoc/Exord12803/Exord12803Page1.html. The United States Conference of Mayors Website: http://www.usmayors.org/uscm/best_practices/private/hawthorn.html. ITT Industries Website: http://www.itt.com/waterbook/where.asp. Investorwords.com: http://www.investorwords.com/2891/long_term_lease.html.

Tax-Exempt Lease

Description: A tax-exempt lease, also called a lease-purchase agreement, is an installment purchase, conditional sale, or lease with an option to purchase for nominal value. In a tax-exempt lease, a public partner finances capital assets or facilities by borrowing funds from a private investor or financial institution. The private partner typically acquires title to the asset at lease signing, but then transfers it to the public partner either at the beginning or the end of the lease term. The lessee is tax-exempt, and thus the lessor is not required to pay federal income taxes on the interest generated by the lease. The issuer of a tax-exempt lease must be a state or possession of the United States, the District of Columbia, or a political subdivision thereof. Nonprofit organizations do not qualify directly to be issuers of tax-exempt obligations, but may be eligible with a sponsoring governmental unit. Personal property such as modular buildings and computers, and real property such as offices and environmental facilities, may be acquired through tax-exempt leases. Tax-exempt leasing is also used for capital expenditures such as implementation of specific projects or expansion of existing facilities.

Reference for Further Information: The National Council for Public-Private Partnerships Website: http://ncppp.org/howpart/ppptypes.shtml. Tax-Exempt Leasing Corporation Website http://www.taxexemptleasing.com/overview.html. Association for Governmental Leasing and Finance Online: http://www.aglf.org/faq.html.

Lease/Purchase

Description: A lease/purchase, also called a tax exempt lease, is a type of installment-purchase contract. Under the lease/purchase model, an organization in the private sector finances and builds a facility which it then leases to a public agency. The facility may be operated by either the public agency or the private organization during the term of the lease. The public agency makes scheduled lease payments to the private organization and accrues equity in the facility with each payment. At the end of the lease term, the public agency either owns the facility or purchases it by paying any remaining unpaid balance in the lease. Lease/purchases enable public agencies to obtain new facilities without relying on capital investment or debt. Lease/purchases have been used for years by the General Services Administration for construction of federal office buildings. The construction of facilities including wastewater and drinking water treatment plants and recycling centers could also be financed through lease/purchases.

Reference for Further Information: See "Tax-Exempt Lease" in this section of the Guidebook. The National Council for Public-Private Partnerships Website: http://ncppp.org/howpart/ppptypes.shtml. Kent, Cheryl, "U.S. Trying Lease-Purchases on its New Buildings," *The New York Times*, May 6, 1990, available at http://query.nytimes.com/gst/fullpage.html?res=9C0CE1DA1530F935A35756C0A966958260.

Sale/Leaseback

Description: A sale/leaseback is a financial arrangement in which the owner of a facility sells it to another entity and subsequently leases it back from the new owner. Public and private entities, including federal, state, and local governments, and private companies, may enter into sale/leaseback arrangements. A tax-exempt lease is a type of sale/leaseback arrangement in which a public entity sells a facility to a private partner in order to finance construction or upgrades, and repays the private partner's investment with lease payments. Another innovative application of the sale/leaseback technique is the sale of a public facility to a public or private holding company for the purposes of limiting governmental liability under environmental statutes. Under that type of arrangement, the public sector partner that sold the facility leases it back and continues to operate it. An example of a sale/leaseback arrangement is Farmer's Insurance Group's sale of its Arizona regional headquarters to Way Commercial Realty in 2007 through a five year, $17.69 million sale/leaseback.

Reference for Further Information: See "Tax-Exempt Lease" in this section of the Guidebook. The National Council for Public-Private Partnerships Website: http://ncppp.org/howpart/ppptypes.shtml. Sorter, Amy Wolff, "Farmer's Insurance Cuts $18 million Sale-Leaseback," GlobeSt.com, September 19, 2007, available at: http://www.cityfeet.com/News/NewsArticle.aspx?PartnerPath=&Id=25963.

Lease/Develop/Operate or Build/Develop/Operate

Description: Under a Lease/Develop/Operate (LDO) or Build/Develop/Operate (BDO) partnership arrangement, a private party leases or buys a facility from a public agency, invests its own capital to renovate, modernize and/or expand the facility, and then operates it under a contract with the public agency. The private partner gets the right to operate the facility for a predetermined length of time and recover its investment through carefully crafted user charges. Roads and bridges, and perhaps municipal transit facilities as well, could be leased and renovated, modernized, or expanded under LDO or BDO partnership arrangements. Virginia has enacted one of the most accommodative public-private partnership laws of any state in the United States, encouraging qualified private sector enterprises to propose to the state transportation department (VDOT) partnership opportunities for investment in new road or transit capacity, increasing the potential for LDO or BDO partnership opportunities in the State. Other states could follow with similar laws. In addition, LDO and BDO arrangements could be used to finance upgrades to environmental facilities, such as wastewater treatment plants and recycling centers.

Reference for Further Information: The National Council for Public-Private Partnerships Website: http://ncppp.org/howpart/ppptypes.shtml.
The Heritage Foundation Website: http://www.heritage.org/Research/SmartGrowth/tst060704a.cfm.

Build-Operate-Transfer

Description: Under the build-operate-transfer (BOT) option, also called build-transfer-operate (BTO) and design-build-operate-maintain (DBOM), a private sector partner builds a facility to the specifications agreed to by a public agency (usually under a turnkey arrangement), operates the facility for a specified time period under a contract or franchise agreement with the agency, and then transfers the facility to the public agency at the end of a specified period of time. Usually the private partner provides some, or all, of the financing for the facility. The build-operate-transfer option is used by communities, municipalities, and other entities for transportation and solid waste management related projects, and for building and upgrading water and wastewater treatment facilities. For example, the Naval Facilities Engineering Command (NAVFAC) Southwest in Camp Pendleton, California is using a DBOM pilot program to enhance operability of its sewage treatment facilities. Through a partnership between the NAVFAC Southwest, the U.S. Marine Corps, and a consulting firm called CDM, this 10-year, $260 million program is delivering facilities that will help to protect the coastal environment from pollution. The program's cornerstone is the 5-million-gallon-per-day Southern Region tertiary treatment plant which applies best technologies for nutrient removal.

Reference for Further Information: The National Council for Public-Private Partnerships Website: http://ncppp.org/howpart/ppptypes.shtml. CDM Website:
http://www.cdm.com/knowledge_center/case_studies/camp_pendleton_water_and_wastewater_improvements.htm.
http://ncppp.org/howpart/ppptypes.shtml. U.S. Department of Transportation Website:
http://www.fhwa.dot.gov/PPP/bot.htm. See "Turnkey" in this section of the Guidebook.

Contract Services: Operations and Maintenance

Description: In contract services carried out for the purposes of operations and maintenance, a public partner, such as a federal, state, or local government agency or authority, contracts with a private partner to provide and/or maintain a specific environmental service or other service. The public partner has the option of retaining ownership and overall management of the public facility or system under this type of contracting arrangement. Under some contract service agreements for operations and maintenance, the risk of operations is shared with the private partner or even transferred to them entirely. Examples of the types of service provided through this type of partnership include lab testing, auditing, the collection of fines and penalties, solid waste collection and disposal, recycling services, asbestos encapsulation or removal operations, the operation and maintenance of water and wastewater treatment facilities and systems, and many other municipal services. For example, Operations Management International, Inc. (OMI) is a water and wastewater service provider that operates and maintains the water and wastewater facilities and equipment for the city of Live Oak, Florida under a contractual agreement.

Reference for Further Information: The National Council for Public-Private Partnerships (NCPPP) Website: http://ncppp.org/howpart/ppptypes.shtml, also see the case study on Live Oak, Florida's Wastewater Treatment Facility on the NCPPP's Website at http://www.ncppp.org/cases/liveoak2.shtml.

Contract Services: Operations, Maintenance, and Management

Description: In contract services carried out for the purposes of operations, maintenance, and management, a public partner, such as a federal, state, or local government agency or authority, contracts with a private partner to operate, maintain, and manage a facility or system providing a public environmental service or other service. Under this contract option, the public partner retains ownership and overall management of the public facility or system, but the private party may invest its own capital in the facility or system. Generally, the longer the contract term, the greater the opportunity for increased private investment because there is more time available for the investor to recoup any investment and earn a reasonable return. Under many operations, maintenance, and management contracts, the risk of operations is shared with the private partner or transferred to them entirely. Local governments sometimes use this type of contract to provide wastewater treatment, solid waste collection and disposal, and recycling services, and other operations and services. For example, the Milwaukee Metropolitan Sewerage District (MMSD) in Milwaukee, Wisconsin has a 10-year contract of this type with United Water for operations, maintenance, and management of the city's municipal wastewater system. This contract enabled the MMSD Commission to reduce user charges by an average of 16.5 percent when it adopted the District's 1999 Operation and Maintenance budget.

Reference for Further Information: The National Council for Public-Private Partnerships Website: http://ncppp.org/howpart/ppptypes.shtml and http://ncppp.org/cases/milwaukee.shtml.

Developer Financing

Description: Under developer financing, a private developer finances the construction and/or expansion of public infrastructure in exchange for the right to build residential housing, commercial stores, and/or industrial facilities served by that public infrastructure. Developer financing arrangements are often called capacity credits, sewer access rights, impact fees, or exactions. The developer financing option is typically under local control; so arrangements can be negotiated on a project-specific basis or mandated through an ordinance. Developer financing is often used for the construction of infrastructure such as sewer lines, biological nutrient removal technology, or whole sewage treatment plants. The Upper Merion Municipal Utility Authority in Pennsylvania uses developer financing through a program requiring customers to pay Sewer Access Rights fees as part of their building, zoning, and mechanical division permit fees. The Town of Dover, Pennsylvania has a similar program.

Reference for Further Information: The National Council for Public-Private Partnerships Website: http://ncppp.org/howpart/ppptypes.shtml. See "Water and Sewer Capacity Credits," "Exactions," and "Impact Fees" in Section 1b of this Guidebook.
University of Maryland Environmental Finance Center Website: http://www.efc.umd.edu/appendixB.html. Upper Merion Township, Pennsylvania Website: http://www.umtownship.org/bc_safety.html. Dover Township, Pennsylvania Website: http://www.dovertownship.org/.

Turnkey

Description: Under a turnkey arrangement, a public agency contracts with a private investor or vendor to design and build a complete facility in accordance with specified performance standards and criteria agreed on between the agency and the vendor. The private developer commits to build the facility for a fixed price and absorbs the construction risks associated with meeting that price commitment. Generally, in a turnkey transaction, the private partner uses fast-track construction techniques, such as design-build, and is not bound by traditional public sector procurement regulations. This combination often enables the private partner to complete facilities in significantly less time and at lower cost than could be accomplished under traditional construction techniques. In a turnkey transaction, financing and ownership of the facility can rest with either the public partner or the private partner. When the private partner provides the financing, it is usually in exchange for a long-term contract to operate the facility. State and local governments use turnkey arrangements to build, operate, and maintain wastewater treatment plants and solid waste disposal facilities. For example, American Water provides the City of Jefferson Parish, Louisiana with full turnkey operation, maintenance, and management of the East Bank Wastewater Treatment Plant.

Reference for Further Information: The National Council for Public-Private Partnerships Website: http://ncppp.org/howpart/ppptypes.shtml. Water Partnership Council Website: http://www.waterpartnership.org/studies/AmericanWater/EastBank.html.

Merchant Facilities

Description: The term "Merchant Facility" refers to a facility run through a public-private partnership where a private sector partner owns and operates the facility, as in privatization deals, and provides services through the facility. Merchant Facilities can be very profitable; but they operate at the whim of market forces. They do not rely on legislated or economic flow control. Flow control is defined by the U.S. Department of Energy as the laws, regulations, and economic incentives or disincentives used by waste managers to direct waste generated in a specific geographic area to a designated landfill, recycling, or waste-to-energy facility. Merchant facilities are often used to provide solid waste management services. They operate as landfills, composting facilities, recycling plants, incinerators, waste-to-energy facilities, and landfill gas power plants. Innovative Energy Systems operates three landfill gas power plants in New York that function as merchant facilities and sell into the day-ahead wholesale energy market.

Reference for Further Information: See "Privatization" in this section of the Guidebook.
U.S. Department of Energy, Energy Information Administration Website, see pp. 8-9:
http://www.eia.doe.gov/cneaf/solar.renewables/renewable.energy.annual/backgrnd/chap7b.htm.
Power Engineering Website:
http://pepei.pennnet.com/display_article/230683/6/ARTCL/none/none/Innovative-Energy-Systems'-Lanldfill-l-Gas-Plants/.

1. Jefferson Parish, Louisiana: Turnkey Operation, Maintenance and Management for a Water System
2. Milwaukee, Wisconsin: Operations, Maintenance, and Management Contract for a Municipal Wastewater System
3. Buffalo, New York: Operations, Maintenance, and Management Contract for a Water System
4. Camp Pendleton, California: Wastewater Treatment
5. Cranston, Rhode Island: Asset Sales Under Executive Order 12803
6. Cartagena, Colombia: Water and Wastewater Services
7. Maryland: Chesapeake Forest Project
8. Morocco: Solar Energy for Rural Households
9. Miami Intermodal Center
10. Lorton, Virginia: South County Secondary School

Jefferson Parish, Louisiana: Turnkey Operation, Maintenance and Management for a Water System

Project Summary: American Water provides full turnkey operation, maintenance and management of the East Bank Wastewater Treatment Plant in Jefferson Parish, Louisiana. This public-private partnership is established through a 15-year contract between American Water and Jefferson Parish that began in 2000 and is priced at approximately $2.5 million annually. The contract provides for five year renewals at the option of Jefferson Parish and it was last renewed in 2005. Under the terms of the contract, American Water provides all dewatering services, offsite sludge transportation, building security, and site landscaping for the treatment plant. In addition, American Water provides odor abatement services at the two major pumping stations using a liquid oxygen injection system.

The East Bank Wastewater Treatment Plant serves approximately 150,000 people in Jefferson Parish. Jefferson Parish has a population of 455,466 and is the largest suburb of New Orleans. The city was affected by Hurricane Katrina in 2005, but has rebounded at a more rapid pace than the neighboring Orleans Parish. A population estimate conducted between June and October 2006 by the Louisiana Recovery Authority following Hurricane Katrina put Jefferson Parish at 440,000 residents or 98% of its 2000 total. Jefferson Parish is successfully rebuilding after Hurricane Katrina.

Hurricane Katrina disrupted American Water's work on an odor control project for the East Bank Plant, but the project is still underway. The company won a $1.4 million design-build-operate proposal for this odor control project. The proposal covers three odor control units for the Belt Press Building and one odor control unit to serve the sludge holding tanks. Phase I of the odor control project was completed in 2001. Phase II of the project, which was postponed due to Hurricane Katrina, is due to be completed between April and June 2008. At completion, the Phase II odor control project will provide full replacement of the facility odor control system and eliminate the overloaded and obsolete Calvert equipment, allowing for greatly improved odor abatement.

Key Achievements and Successes: The citizens of Jefferson Parish are enjoying high quality wastewater treatment plant operations and an annual cost savings close to $1.56 million. American Water made the following improvements to the treatment plant during the first five years of the contract: 1.) Installation of a state-of-the art, plant-wide Supervisory Control Data Acquisition (SCADA) system for full monitoring and control of the entire facility from a central console, 2.) Installation of a new and fully configured Computerized Maintenance Management System to document the plant's assets, 3.) Installation of a new effluent pumping control system that saves energy in plant operations and guards against effluent overflows during the extreme wet weather flows that are common in southeast Louisiana.

Reference for Further Information: Water Partnership Council Website: http://www.waterpartnership.org/studies/AmericanWater/EastBank.html.
American Water Website: http://www.amwater.com/awpr1/default.html, American Water phone: 504-736-6299.

Milwaukee, Wisconsin: Operations, Maintenance, and Management Contract for a Municipal Wastewater System

Project Summary: Great savings have been realized through the ten year contract between the Milwaukee Metropolitan Sewerage District (MMSD) in Milwaukee, Wisconsin and United Water Services (UWS), which was set up to provide for the operations, maintenance, and management of the MMSD's municipal water system. This contract is the largest wastewater public-private partnership agreement in the United States, and it is a 1999 National Council for Public-Private Partnerships project award winner. Milwaukee, with a population of 573,378 as of the 2006 United States Census, is the largest city within the state of Wisconsin. The partnership has already begun paying dividends for Milwaukee-area residents and has set a new standard for municipalities across the country considering public-private partnerships. This contract enabled the MMSD Commission to reduce user charges by an average of 16.5 percent when it adopted the District's 1999 Operation and Maintenance budget.

The MMSD is a special purpose municipal corporation organized under the laws of the State of Wisconsin; and its legal boundaries include all of Milwaukee County with the exclusion of the City of South Milwaukee and small areas in the City of Franklin. The legal boundaries also include the portions of the Village of Bayside and the City of Milwaukee that are in neighboring counties. The MMSD provides sewage treatment services for the 18 cities and villages within its legal boundaries. In addition, the MMSD is authorized under state statutes to provide sewage treatment services for areas beyond its legal boundaries but within the portion of the multi-county drainage basin delineated as part of the Water Quality Management Plan developed by the Southeastern Wisconsin Regional Planning Commission, an area including all or parts of 10 municipalities outside Milwaukee County.

Key Achievements and Successes: The contract between MMSD and UWS has some unique features and notable achievements associated with it. First, it includes a no layoff guarantee from UWS for the entire term of the contract, the first of its kind to be included in a competitive contract. Second, it includes a pension agreement. The U.S. Internal Revenue Service and the U.S. Department of Labor ruled in May 1999 that the MMSD, United Water Services and the city of Milwaukee can remain in the City of Milwaukee's public employee pension fund without compromising its status as a government plan.

Reference for Further Information: National Council for Public-Private Partnerships Website: http://ncppp.org/cases/milwaukee.shtml. United Water Website: http://www.unitedwater.com/municpal.htm, scroll down and click on "Milwaukee, Wisconsin." Milwaukee Metropolitan Sewerage District Website: http://www.mmsd.com/home/index.cfm.

Buffalo, New York: Operations, Maintenance, and Management Contract for a Water System

Project Summary: In September 1997, the Buffalo, New York Water Board entered into a contract with American Water Services, Inc. to upgrade, operate, and maintain its water system. This public-private partnership is a 2005 National Council for Public-Private Partnerships Service Award winner. The original contract had a five-year term with a one year extension. The contract has since been renewed for another five years beginning July 1, 2003. This partnership between American Water and the City of Buffalo has made significant improvements to the city's water system, including the complete automation of customer records and general operations and the design and construction of a new state-of-the-art customer service center with easy access to mass transit.

The services American Water provides to the City of Buffalo include repair and maintenance of the water distribution system, water treatment and pump station operation, residuals management, customer service, billing and collections, and the repair and installation of water meters. These services are benefiting a substantial population- Buffalo is the State of New York's second largest city after New York City, with a population of 292,648. Buffalo is the economic and cultural center of the Buffalo-Niagara Falls metropolitan area which has a population of 1.2 million. The Buffalo water system has 80,000 connections serving the city's entire population and drawing its water from Lake Erie with an average flow of 78-91 million gallons per day.

Key Achievements and Successes: This public-private partnership provides the following benefits to the City of Buffalo and its residents and utility customers:

- The City of Buffalo has saved $21 million through operational and financial management improvements to its water system over six years of working with American Water.
- An initial water rate reduction of 8% was held for five years.
- The innovative labor contract utilizes city employees with no involuntary staff reductions.
- Changes in work rules and improved deployment yield a 26% increase in productivity for city employees working under this partnership.
- The collection rate for the water system increased from 80% range to 97% range, resulting in significant positive revenue impact.
- The water system converted from flat rate to metered water, with installation of over 60,000 water meters.
- Improvements in water quality achieved through best practices reduced turbidity by 450%.
- Responsiveness and efficiency of water line repairs improved significantly with implementation of the computerized maintenance and management system (CMMS).
- Vehicle reliability for the water system improved via a new replacement and repair program. The average age for vehicles in the fleet was reduced from 14 years to 8 years.

Reference for Further Information: The National Council for Public-Private Partnerships Website: http://www.ncppp.org/cases/buffalo.shtml. Water Partnership Council Website: http://waterpartnership.org/studies/AmericanWater/Buffalo.html.

Camp Pendleton, California: Wastewater Treatment

Project Summary: The Naval Facilities Engineering Command (NAVFAC) Southwest, based in San Diego, California, is using a design-build-operate-maintain (DBOM) pilot program to enhance operability of its sewage treatment facilities in Camp Pendleton, California through privatization. Camp Pendleton is the major west coast base of the United States Marine Corps, serving as its prime amphibious training base. The base is located in Southern California between the cities of Oceanside and San Clemente. Camp Pendleton has a daytime population of 60,000 and maintains 7,300 family housing units.

Through a partnership between the NAVFAC Southwest, the U.S. Marine Corps, and a consulting firm based in Carlsbad, California called CDM Constructors, Inc., this 10-year, $260 million program is delivering facilities that will help to protect the coastal environment from pollution. CDM Constructors, Inc. has been awarded a firm-fixed price, indefinite-delivery/indefinite-quantity DBOM contract for water conveyance/reclamation at Camp Pendleton. The NAVFAC Southwest is responsible for the contracting activity.

The program's cornerstone is the new 5-million-gallon-per-day Southern Region Tertiary Treatment Plant that consolidates five outdated sewage treatment facilities and applies best technologies for nutrient removal. The work to be performed provides for the installation of a conveyance/reclamation pipeline to connect four sewage treatment plants to the Southern Region Tertiary Treatment Plant to ensure compliance with Clean Water Act requirements. This work is expected to be completed by March 2008.

Key Achievements and Successes: This partnership is delivering flexible, reliable, safe, and easy-to-operate facilities that will help Camp Pendleton to meet rigorous water quality compliance objectives under the Clean Water Act. The creative collaboration and communication approaches integrate multidisciplinary stakeholder interests with CDM's technical guidance. The program is also delivering advanced treatment processes that meet significant physical and procedural site restrictions, balancing technical and scheduling priorities with environmental compliance.

Reference for Further Information: CDM Website:
http://www.cdm.com/knowledge_center/case_studies/camp_pendleton_water_and_wastewater_improvements.htm.
U.S. Environmental Protection Agency Website: http://www.epa.gov/fedrgstr/EPA-IMPACT/2001/April/Day-25/i10221.htm and http://www.epa.gov/fedrgstr/EPA-IMPACT/2004/June/Day-22/i14107.htm.
Defense Industry Daily Website: http://www.defenseindustrydaily.com/sewage-infrastructure-at-camp-pendleton-ca-prepares-for-privatization-01958/.
See "Build-Operate-Transfer" in Section 4a of this Guidebook. Build-operate-transfer is another name for design-build-operate-maintain.

Description: Executive Order 12803, issued in 1992, directs all United States federal departments and agencies to approve state and local governments' requests to privatize infrastructure assets financed in whole or part by the federal government to the extent permitted by law and consistent with originally authorized purposes. An example of a contract approved under Executive Order 12803 is Cranston, Rhode Island's lease of its entire wastewater treatment, collection, and pumping system to Poseidon Resources Corporation of Stamford, Connecticut for 25 years starting in 1997. This long-term lease arrangement is valued at $400 million, making it the second largest public-private partnership for municipal water or wastewater treatment in the United States. Cranston is the third largest city in Rhode Island, with a population of 79,269 as of the 2000 Census.

The City of Cranston issued an RFP in 1997 for the purchase of its 23 gallons per day (mgd) wastewater treatment plant. The city was unable to select a firm based upon the bids received and chose to enter into competitive negotiations with several firms instead. An employee-owned architectural, engineering and consulting firm, HDR, was brought in to assist the city with the competitive negotiation process. The city chose to lease the wastewater treatment plant for a 25-year term to Poseidon rather than sell it. This contractual lease arrangement was approved by the Cranston City Council and signed in March 1997. The lease contract is set up to provide an innovative solution to meet all of Cranston's intermediate and future wastewater objectives.

In addition to Poseidon, there are a couple of other private organizations in this lease arrangement. Professional Services Group, Inc. (PSG), of Houston, Texas, who has been operating, maintaining, and managing Cranston's wastewater treatment plant since 1989, will continue to provide those services for the whole system. Metcalf & Eddy, Inc., (M&E), of Branchburg, New Jersey will provide a treatment facility upgrade in addition to design/build services for capital improvements. This contract is the first in the United States where the contractor has full financial responsibility for a wastewater system over the term of the service agreement. Still, the lease arrangement allows Cranston to regain control of the treatment facility asset at the end of the 25-year agreement.

As part of the contract, Cranston received a cash payment of $48 million from Poseidon. That payment will be used by the city to retire general obligation debt and repay money borrowed from the city's general fund. Those transactions will remove approximately one third of Cranston's outstanding debt and enhance the city's credit rating. In addition, part of the $48 million payment will be used to provide funds for Cranston's wastewater system upgrade and generate excess cash for the city's sewer enterprise fund. Out of the $48 million transaction, $24 million is dedicated towards initiatives to help the city to comply with environmental regulations. The compliance made possible by that financing could lead to an improved agency rating that would yield future financial benefits to the city.

Key Achievements and Successes: The partnership created by this lease arrangement provides great environmental protection benefits and financial benefits for the City of Cranston and is beneficial for the consumers and the private contractors as well. The contract is expected to save the City of Cranston $74 million. As far as the consumer is concerned, this arrangement means that there will be no immediate increase in user rates and there will be an increased likelihood of stable, predictable rates for the next 25 years. The lease is expected to provide the members of Professional Services Group's operating and renovation team with opportunities to expand their businesses. Poseidon is earning equity through its leasing and financing. Metcalf & Eddy is expanding its potential to take advantage of design/build opportunities through its participation in the contract. This public-private partnership creates a win-win situation for all participants and the consumer as well. This very successful partnership can serve as a model for other cities.

Reference for Further Information: Executive Order 12803 as of April 30, 1992, available at http://www.theantechamber.net/UsHistDoc/Exord12803/Exord12803Page1.html. See "Asset Sales Under Executive Order 12803" in this section of the Guidebook.
Water & Wastes Digest Website: http://www.wwdmag.com/Long-Term-Lease-of-Treatment-Systems-Becomes-an-Option-article882.
HDR Website: http://www.hdr-engineering.com/13/38/1/default.aspx?projectID=608.
ITT Industries Website: http://www.itt.com/waterbook/where.asp and http://www.itt.com/waterbook/privatization.asp.

Project Summary: Cartagena is a large city seaport on the northern coast of Colombia with a population of 895,400. Colombia is a leader in policies for improved conditions in water supply and sewerage. Years of political pressure in Cartagena led to the municipal council's action creating a public-private partnership to repair and operate the city's water and wastewater system in 1994. The municipal council, under the leadership of the Mayor, approved the creation of an "empresa mixta" (mixed enterprise) which combined the resources of the city's public works department and a Spanish water firm. This public-private partnership operates under the name Aguas de Cartagena (AGUACAR). Under the AGUACAR partnership, the shares of the Spanish water company were split, with the municipality owning 50%, the Spanish water firm owning 46%, and other private investors owning 4%. The Municipality of Cartagena granted AGUACAR a 26-year concession contract to operate and maintain water supply and sanitation services, collect water tariffs, and enforce payment.

The project AGUACAR has carried out is called the Cartagena Water Supply, Sewage, and Environmental Management Project. The project investment, including grants, loans, and the initial expenditures to organize AGUACAR, totaled US $117.2 million. The initial cash investment in AGUACAR was US $4.6 million. The private sector contributed 50% of that capital. After the initial investment, additional funding for the project was obtained from the Colombia national government (US $20 million) and the municipality of Cartagena (US $7.6 million). In 1999, AGUACAR was given an US $85 million International Bank for Reconstruction and Development (IBRD) loan through the World Bank to help finance sewerage, wastewater treatment facilities, and water supply infrastructure. The World Bank has provided a series of loans for water and sewer system improvements, totaling more than US $700 million to Colombia since 1988. Aguas de Cartagena's project was the first of these World Bank funded projects that included a public-private partnership.

The AGUACAR partnership had several important objectives for this project. The primary objective was to provide reliable water and wastewater services to all of Cartagena's residents, especially lower income households. Before this partnership was created, the lower income portions of the city were not connected to the municipal water system. Customers that were connected to the system found that water pressure was generally low and often interrupted. A related objective of AGUACAR was to decrease the amount of Non-Revenue Water (NRW) caused by leaks in the existing pipes. The city was losing between 40% and 60% of its water through leaks and its goal was to reduce this loss to no more than 25%. A third objective of AGUACAR was to improve the economics of the water system, which was operating at a substantial loss, and make it a financially self sustaining operation. This goal was to be achieved through increased efficiency in the management of water supply and sewage services and an increase in the tariff collection ratio.

Key Achievements and Successes: Aguas de Cartagena has been very successful in achieving its objectives. Through the achievement of these objectives, the financial health of Cartagena's water system has been improved and access to water services for the city's residents has greatly increased. Soon after the start of the partnership, AGUACAR began immediate service to those not connected to the water system through a system of water truck deliveries. The financing for those deliveries came in part from a restructuring of the tariff system to incorporate cross-subsidies, under which more affluent customers helped subsidize the water rates for lower income families. Next, to expand service, about 35,000 new connections to the water system were made, almost exclusively in poor neighborhoods. By 2005, through the work of the AGUACAR partnership, water system connections in Cartagena increased to 99% of the population and sewage coverage rose to 75%.

Another great accomplishment of AGUACAR was that the percentage of Non-Revenue Water (NRW) caused by leaks in Cartagena's system was significantly reduced through the partnership's work. As a result, customers began to receive more reliable water pressure. Pipelines and water meters were evaluated and repairs were made to achieve this reduction in leaks. These repairs meant that the quantity of the overall water supply and the amount of time each day that water was available to customers was greatly increased. In addition, the long-term financial situation of the water system was improved as compared to the many years when the city lost a lot of money by treating and distributing water that was lost before reaching consumers. Still, this improvement in the financial health of the city's water system was not adequate by itself. The city had a low rate of water bill collection so AGUACAR worked to make some improvements in that area as well.

The AGUACAR partnership made the goal of increasing the collection of fees ratio for Cartagena's water provider from less than 50% to 95%. Many households lacked working water meters before AGUACAR began its work. That made it very difficult for the water provider to collect tariffs and contributed to negative cash flows. To fix this problem, AGUACAR installed or replaced water meters at all new and existing water system connections. By 2005, over 99% of connected households had water meters. Through improved billing and tariff collection, made possible in part by the installation of the water meters, the water provider's collection of fees ratio was greatly improved, reaching 90% in 2005. Increases in efficiency of water distribution coupled with a better tariff collection record gave AGUACAR positive cash flows. Support for AGUACAR is high, as Cartagena's utility customers benefit from improvements in water and sewerage services that would not have been financially possible without the partnership.

Reference for Further Information: The National Council for Public-Private Partnerships Website: http://www.ncppp.org/undp/cartagena.html.

Maryland: Chesapeake Forest Project

Project Summary: The Chesapeake Bay is the largest estuary in the United States; and it is a major area for both recreational and commercial fishing in the State of Maryland. It lies off the Atlantic Ocean and is bordered by Maryland, Virginia, the District of Colombia, New York, Pennsylvania, Delaware, and West Virginia. The environmental quality of the Bay has been seriously degraded by wastewater discharges from growing population centers and agricultural runoff in recent decades. In response to this environmental degradation, governments in the Chesapeake Bay watershed have made restoration of the Bay an environmental priority, and have shown an increased interest in land and wetlands management.

However, the state and local governments in the Bay's watershed lack the financial and personnel capability to address many of these environmental problems adequately. It was these types of financial shortfalls that inspired the creation of the Chesapeake Forest Project (CFP), a public-private partnership in Chesapeake Forest, Maryland. The CFP was initiated in 1999 when the State of Maryland entered into a public-private partnership to purchase 58,000 acres of land owned by a private lumber company in the forested areas of the Eastern Shore of the Chesapeake Bay.

The State of Maryland Department of Natural Resources (DNR) is the public agency with the direct responsibility for oversight and management of all phases of the CFP. A major philanthropic foundation aided in the initial acquisition of a portion of the property for the CFP. A non-profit public interest group with its primary purpose being environmental protection is another participant in the development and operation of the CFP. The third and final partner is a for-profit consulting forestry firm that performs the forestry management under the implemented CFP.

For the initial land acquisition that ultimately created the partnership, the State of Maryland provided $16.5 million for the purchase of half of the 58,000 acres. The non-profit public interest group, acting on behalf of the philanthropic foundation, purchased the remaining 29,000 acres for $16.5 million with plans to later gift the land to the State of Maryland. The gifting, done in December 2000, carried with it a number of stipulations about terms that would later become key parts of the Chesapeake Forest Project agreement.

Stipulations of the non-profit group's gifting of the land included commissioning of a detailed and comprehensive Sustainable Forest Management Plan, to be implemented over a three-year transition period. The for-profit consulting forestry firm was contracted to manage the Chesapeake Forest lands in conformance with the State of Maryland's environmental standards and regulations and based upon the Sustainable Forest Management Plan. The contract provided the consulting forestry firm with the option of participating in the final partnership agreement, when implemented.

Under this partnership, the for-profit consulting forestry firm is responsible for all land management, based on an annual per-acre fee for basic management services, including harvesting timber. The forestry firm pays all subcontractor bills and is responsible for subcontracting all fieldwork, including tree cutting, transportation of wood products to the mill, and replanting. Receipts from the sale of timber products are paid directly to the State of Maryland and the state does the financial management of the accounts. Each year, the first 15% of the revenues are dispersed to the local counties. The forestry firm assumes substantial risks in that only after the payment to local counties has been made may it be paid its management fee.

The broad vision for the Chesapeake Forest Project is the creation of an active, working, certified sustainable forestry initiative on the Eastern Shore of the Chesapeake Bay that pays for itself and supports local communities. The State of Maryland and its private partners in the CFP seek to achieve the following objectives:

● Provide employment and a steady flow of economic activity to support local businesses and communities;

● Prevent the conversion of forested lands to non-forested land uses;

● Contribute to water quality improvements as part of the larger Chesapeake Bay restoration effort;

● Protect and improve habitat for threatened and endangered species,

● Help sustain soil and forest productivity and health; and,

● Protect sites of special ecological, cultural, and historical interest.

Key Achievements and Successes: Possible political resistance to this project was reduced through compliance with the State of Maryland's laws that required the financial management of the project to be done by a state agency, thus ensuring transparency from initiation to final implementation of the partnership. A unique aspect of this public-private partnership is that it is self-funded. The Sustainable Forest Management Plan includes identification of the areas in the forest where wood products can be harvested at an environmentally sustainable level without negative environmental impacts. The revenues from the harvesting pay for the contract and provide additional funds to state and local governments. In addition, the controlled continuation of timber harvesting activities has addressed the economic concerns of local communities. This partnership has successfully established a state-owned public forest, managed on a daily basis by a private forestry firm, carrying out a conservation-oriented sustainable forestry plan.

Reference for Further Information: The National Council for Public-Private Partnerships Website: http://ncppp.org/undp/chesapeake.html.

Morocco: Solar Energy for Rural Households

Project Summary: Morocco's Office National de l'Electricite (ONE), the state-run operator in Morocco's electricity supply sector, entered into a public-private partnership in June 2002 with a private company to electrify rural households by using solar energy to produce electricity in an initiative called the rural electrification project. The private partner is the Renewable Energy Service Company (RESCO), which is comprised of a French oil company, a French electricity company, and one of their joint subsidiaries which provides design, production, installation and operation of photovoltaic power systems. The solar customers become clients of ONE once they sign the utility contract. However, RESCO is responsible for the installation and maintenance of solar equipment as well as the collection of users' fees in 24 of Morocco's 62 prefectures and provinces.

The primary objective of the rural electrification project is to provide photovoltaic kits to over 58,000 households in rural Morocco to enable them to meet their basic electricity needs. Based on the 2004 census, the population of Morocco is 29.7 million, and 45% of that population lives in rural areas. Many rural residents in Morocco are not connected with the electric grid. The Moroccan government maintains that incorporating rural households in to the electric grid is too costly a venture. Electricity demand is unevenly distributed throughout the country and the distance between homes and from homes to the grid makes it difficult to incorporate them into the grid. Fortunately, using solar power to provide rural homes with electricity outside the grid is a viable option for Moroccans and the rural electrification project is proving to be a great success in the areas of reliability, affordability, and financing.

The total investment budget of the rural electrification project is $35.5 million, with 66% of those dollars being provided by an equipment grant from ONE. The equipment grant was largely financed through a $6.5 million grant from the German Bank KfW, a $6.5 million soft loan from the French Development Agency (AFD), and a $1.5 million grant from the French Fund for the World Environment (FFEM). The start-up phase to provide technical assistance for the project was financed with the equipment grant. An equipment subsidy is provided by ONE, enabling the partnership to provide electrical services at affordable rates, by offsetting the high installation and maintenance costs associated with solar home systems. This subsidy enables the partnership to offer rural customers electric rates that are adapted to household budgets. It also offers rates that are comparable to what Moroccan households, that are connected to the grid, pay for service.

Key Achievements and Successes: The private partner (RESCO) delivered prompt and reliable service while keeping costs low through the hiring and training of local technicians for the rural electrification project. By July 2005, more than 14,000 households in 400 villages had received solar home systems through the project, with an installation rate of 500-700 homes per month. The 16,000 customers of the first phase of the project were connected before the end of 2005, one year in advance of the contract schedule. The second phase started in the second quarter of 2005. The installation of solar home systems for the remaining 37,000 customers is scheduled to take two years. The success of this project has helped to demonstrate that a public-private partnership can provide affordable solar electricity and jobs to people living in rural areas of Morocco.

The hiring and training of local technicians for the project provided jobs in areas where unemployment is high and enabled RESCO to offer its customers rates that are similar to what they previously paid for candles, gas, and batteries or battery recharging. In addition, RESCO developed a reputation for accessibility and trustworthiness throughout the communities by having local offices and local representation at its weekly meetings. That attention to customer support, combined with the affordable electric rates, has helped RESCO to maintain a low payment default rate. This public-private partnership's success in providing solar electricity to rural communities in Morocco is a great achievement in the areas of environmental protection and rural development that could be looked at as a model by other countries with similar needs and demographics.

Reference for Further Information: The National Council for Public-Private Partnerships Website: http://www.ncppp.org/undp/morocco.html.

Miami Intermodal Center

Project Summary: The Miami Intermodal Center (MIC), scheduled to be completed in May 2011, is a very large ground transportation hub including a train station that is being developed through the MIC Program by the Florida Department of Transportation (FDOT) in Miami-Dade County, Florida. The MIC Program was created in the summer of 1993 when FDOT entered into a partnership with six federal agencies of the U.S. Department of Transportation. In May 1998, Kaiser Engineers was awarded the consultant program manager contract for the MIC Program. Kaiser and its subconsultants formed a consulting management team to assist the FDOT with the planning, design, and implementation of the MIC Program. Recognizing the need for inter-agency cooperation and coordination, FDOT entered into strategic partnerships in 2000 with Miami-Dade County, Miami-Dade Expressway Authority, and what is now the South Florida Regional Transportation Authority.

Another MIC partner is Earth Tech Inc., a business unit of Tyco International Ltd., and a global provider of consulting, architectural, engineering and construction services. Earth Tech has been working on the MIC starting with the initial MIC Program development in 1993. The company is currently serving as MIC Program manager for the public side of the project, which has been recognized by the U.S. Department of Transportation as a "Project of National Significance." In 2007, it was announced that Earth Tech will be on board for Phase One of Miami Central Station. District Six of FDOT awarded Earth Tech a $3 million, 15-month contract to plan, design, and prepare construction documents for the facility. Miami Central Station will consolidate Tri-Rail and Amtrak passenger rail services under one roof. The station design accommodates a potential Florida High Speed Rail Terminus.

The MIC is being constructed next to Miami International Airport. The origin of the MIC can be found in reports dating back to the 1980s when county officials foresaw the need to create connectivity for the region as Miami-Dade began to attract big businesses and experience more substantial population growth. The county's population was 2,253,362 at the time of the 2000 Census. The U.S. Census Bureau estimates the 2006 population of Miami-Dade County at 2,402,208. When completed, the MIC will include the Miami Central Station rail station, roadway improvements, and a Rental Car Center to serve this growing population. It will ease traffic congestion by providing connections where none previously existed between all forms of ground transportation in the Palm Beaches, Fort Lauderdale, Miami, and the Florida Keys. Much work has been done to coordinate the financing that makes this project possible.

The cost of the MIC is $1.349 billion. The MIC Program is the largest ever surface transportation investment by the federal government. The Transportation Infrastructure Finance and Innovation Act (TIFIA) was passed on June 9, 1998 as a federal credit program backed by leveraging federal funds to attract private investment in transportation infrastructure. Federal support for intermodal projects such as the MIC Program was authorized under TIFIA. In 1999, the MIC Program was selected to receive up to $439 million in two separate authorizations through TIFIA. The first loan of $269 million was closed on June 9, 2000, but only $15 million of it was withdrawn because FDOT replaced it with a more competitive internal loan through the State Transportation Trust Fund. The second loan of up to $170 million was closed on April 29, 2005, and an additional $100 million has been requested to augment the second loan.

Continuation of MIC funding was made possible by legislation passed in 2005. In August 2005, the Safe, Accountable, Flexible, Efficient Transportation Equity Act: A Legacy for Users (SAFETEA-LU), a 6-year federal transportation bill, was passed to continue funding of federal surface transportation programs that were funded under TEA-21. This legislation gives state and local transportation decision makers more flexibility for solving transportation problems in their communities. The MIC Program is included in SAFETEA-LU. This continuation of funding means that there is great promise for the MIC to be completed successfully. The MIC will provide convenience and environmental protection benefits by expanding access to energy efficient train travel and making the use of train travel more convenient in Miami-Dade County and the South Florida region.

Key Achievements and Successes: Substantial progress is being made on the construction projects of the MIC program. All of the MIC roadway improvement projects have either been completed or are under construction. Construction of the Rental Car Center began in the spring of 2007. Design work on the Miami Central Station is nearing completion. A site has been chosen for it north of the existing Tri-Rail station. The MIC Rail Station Phase One, a component of Miami Central Station, received an award in the "Unbuilt" category from the Miami Chapter of the American Institute of Architects (AIA). The honors were presented to the MIC design team at the 50th Annual AIA Design Awards Gala in November 2004.

This AIA award acknowledges the MIC design team's commitment to make the MIC a landmark entrance to Miami, thus fulfilling the promise of State Secretary Jose Abreu to deliver structures that are as efficient as they are appealing. In comments made in 1998 to the Miami-Dade County Metropolitan Planning Organization, Secretary Abreu said the MIC would be Miami's gateway because 95% of Miami-Dade county's visitors arrive through Miami International Airport and either rent cars or take other modes of transportation to destinations within South Florida. The MIC is expected to accommodate South Florida's expansion, trigger economic development, and enhance the quality of life in the state.

Reference for Further Information: Miami Intermodal Center Website:
http://www.micdot.com.
Earth Tech Website:
http://www.earthtech.com/news/EarthTechMiamiIntermodalCenterMICCentralStationDesign.htm.
U.S. Census Bureau Website: http://www.census.gov/, search on Miami, Florida.
InnovativeFinance.org: http://www.innovativefinance.org/projects/intermodal/miami.asp.

Lorton, Virginia:
South County Secondary School

Project Summary: It became clear in 2001, years before the budgeted funding would be available, that a new secondary school was needed in southern Fairfax County, Virginia. The lengthy commutes faced by students were a serious hardship. An innovative finance solution was clearly necessary for the new school because the Public-Private Educational Facilities and Infrastructure Act of 2002 had not yet been passed. Fairfax County Public Schools solicited proposals in 2001 for public-private partnerships for the construction of a new school, to be called South County Secondary School. A partnership between Clark Education, a private entity, and Fairfax County Public Schools, a government entity, was established for the development, design, financing and construction of the school on a portion of the former prison site in Lorton, Virginia. Fairfax County Public Schools was the first K-12 public school system in Virginia to try this alternative, turnkey approach to procuring a school.

Clark Education provided financing solutions including monetization of land and value engineering that allowed the school to be built three years ahead of schedule. Value engineering is a specialized cost-control technique that uses a systematic and creative approach to identify and reduce unjustifiably high costs in a project without sacrificing the reliability or efficiency of the project. Clark worked successfully with the school's architect to reduce unjustifiably high construction costs. In addition, Clark's monetization of specific parcels of unused Fairfax County land assets lowered the overall net cost of the school project to the county. The school was completed as scheduled in August 2005, years before financing would have been made available if the public-private partnership had not been established. In September 2005, South County Secondary School opened and hosted the first group of students to attend the school.

Key Achievements and Successes: This partnership is a 2006 National Council for Public-Private Partnerships Innovation Award winner. Clark delivered the school at a savings of just over $25 million against the $88.7 million project budget without diverting resources away from other Fairfax County school projects. These savings freed up funds that could potentially be used for the construction and maintenance of other public and private infrastructure in Fairfax County, such as water and wastewater treatment plants. This state-of the-art school is greatly appreciated by the community. Although South County was built based on designs of a neighboring high school, it has more natural light in the classrooms and partitions that were designed to control noise above and below the ceilings. The school's technology, including over 800 computers and a two-story media center, exceeds that of any previously constructed school in Northern Virginia.

Reference for Further Information: The National Council for Public-Private Partnerships Website: http://www.ncppp.org/cases/southcounty.shtml. See "Value Engineering" in Section 3 of this Guidebook and "Turnkey" in Section 4 of this Guidebook. See the "turnkey" definition on The National Council for Public-Private Partnerships Website: http://ncppp.org/howpart/ppptypes.shtml.

Financial outreach has proven to be very important for community environmental programs and small businesses because of the multiplicity and complexity of ever increasing environmental regulations, both at the federal and state level, and the need to finance, operate, improve, or construct facilities to comply with these regulations. Financial outreach is a critical link between environmental mandates and implementation of these mandates by local managers.

Two types of financial outreach tools are presented in this section: **institutional arrangements and electronic services.** The institutional arrangements presented include organizations, initiatives, and mechanisms that support the financing of environmental programs, systems, and projects. The electronic services presented include Websites that provide information on funding mechanisms and software programs that are tools for financial planning.

The institutional arrangements in this section tend to be independent, innovative, and non-bureaucratic. They typically involve face-to-face, hands-on training, and are project specific. They often encourage significant client involvement ranging from detailed feedback and cooperation to direct project participation and funding. As a result, the quality of the outreach and technical assistance they provide is enhanced and their services are financially leveraged.

Using electronic services, people can interact and access information in ways that are often very cost effective and save time as well. Properly implemented, electronic services can help control resource consumption and pollution by reducing paper use, cutting transportation and fuel costs, and thus preventing related air, water and land pollution. In addition to the electronic services described in this section, some electronic services for businesses are discussed in Section 10 of the Guidebook.

1. U.S. Environmental Protection Agency: Environmental Finance Program
2. Environmental Finance Center (EFC) Network
3. Region 1 Environmental Finance Center at the University of Maine
4. Region 2 Environmental Finance Center at the Maxwell School, Syracuse University
5. Region 3 Environmental Finance Center at the University of Maryland
6. Region 4 Environmental Finance Center at the University of North Carolina
7. Region 4 Environmental Finance Center at the University of Louisville
8. Region 5 Environmental Finance Center at Cleveland State University
9. Region 6 Environmental Finance Center at the University of New Mexico
10. Region 9 Environmental Finance Center at Dominican University
11. Region 10 Environmental Finance Center at Boise State University
12. Boise State University Environmental Finance Center: Plan2Fund
13. Boise State University Environmental Finance Center: CAP Finance
14. Finance Charrettes
15. Border Environmental Cooperation Commission
16. Rural Community Assistance Corporation
17. Self-Help
18. Circuit Riders
19. Cooperatives
20. Drinking Water State Revolving Fund Capacity Development
21. U.S. Environmental Protection Agency: Water and Wastewater Pricing Website
22. U.S. Environmental Protection Agency: Enforcement Economic Models
23. U.S. Environmental Protection Agency: Green Power Locator
24. U.S. Department of Energy: Financial Opportunities Website
25. U.S. Department of Energy Alternative Fuels Data Center: State and Federal Incentives and Laws Website
26. National Technical Assistance Programs
27. National Rural Water Association
28. Rocky Mountain Institute
29. U.S. Environmental Protection Agency: Water Efficiency Market Enhancement Program
30. Environmental Council of the States
31. Pollution Prevention Resource Exchange
32. Energy Efficiency Organizations
33. U.S. Department of Energy: Industrial Assessment Centers
34. Catalogue of Federal Domestic Assistance
35. Database of State Incentives for Renewable Energy (DSIRE)
36. U.S. Environmental Protection Agency: Catalog of Federal Funding Sources for Watershed Protection
37. Boise State University Environmental Finance Center: Directory of Watershed Resources

U.S. Environmental Protection Agency: Environmental Finance Program

Description: The U.S. Environmental Protection Agency (EPA) Environmental Finance Program works in partnership with state and local governments, tribes, and the private sector to help fund environmental protection initiatives. The Program provides leveraged financial outreach services to these partners through three distinct, but related components: the Environmental Financial Advisory Board (EFAB), the Environmental Finance Center Network (EFCN), and the Environmental Financing Information Network (EFIN).

EFAB, a federally chartered advisory committee, provides innovative ideas and recommendations to the EPA on ways to lower the costs of, increase investments in, and promote public-private partnerships with regard to environmental and public health protection. The EFC Network, consisting of nine university-based programs that serves eight of the ten EPA regions, delivers targeted technical assistance to address the "how to pay" issues of meeting environmental standards. EFIN catalogues the work and accomplishments of EFAB and the EFCN, and provides abstracts of valuable Environmental Finance Program publications, and some EPA publications, through its highly popular Website, telephone Infoline, and contact referral service.

Reference for Further Information: Environmental Finance Program Website: http://www.epa.gov/efinpage, EFIN Infoline: 202-564-4994. The EFC Network is described in this section of the Guidebook.

Environmental Finance Center (EFC) Network

Description: The Environmental Finance Center Network (EFCN) is a system of nine university-based Regional Environmental Finance Centers that provide state and local governments and the private sector with training and educational, technical, and analytic assistance on environmental finance (see the individual descriptions of the nine Environmental Finance Centers on the following pages). These services are designed around the "how to pay" issues of environmental compliance. The EFCN has become a significant force in assisting local governments and small businesses in meeting environmental standards. A central goal of the Network is to help create sustainable environmental systems in the public and private sectors. Coordination of the EFC Network is provided by the U.S. Environmental Protection Agency (EPA) Environmental Finance Program. In addition to the nine mature Centers, in 2008 EPA's Region 7 office created a satellite EFC with the intention of moving it into the Network within a year.

Reference for Further Information: Contact: Vera Hannigan, E-mail hannigan.vera@.epa.gov, Phone # 202-564-5001. EFC Network information and a Website for each EFC can be accessed via the Environmental Finance Program's Website at http://www.epa.gov/efinpage. The Environmental Finance Program is described in this section of the Guidebook.

Region 1 Environmental Finance Center at the University of Southern Maine

Description: Founded in 2001, the University of Southern Maine EFC – or the New England EFC (NE/EFC) – is housed within Region 1 in the Edmund S. Muskie School of Public Service. The NE/EFC addresses the "how-to-pay" questions associated with creative approaches to environmental protection and management, especially by developing and applying "smart growth" and other land-use techniques that go beyond compliance with government regulations.

In developing its programs, the NE/EFC identifies the research, education, and technical assistance needs of an array of clients from land trusts, developers, and municipalities, to state governments and agencies of the federal government. The NE/EFC offers Collaborative Environmental Services tailored to the needs of individual organizations or local governments. These include assistance in facilitating local dialogues and exploration of creative ways to make needed conservation and development decisions feasible financially and otherwise. The EFC has also developed a program called the Next Communities Initiative, to train community leaders how to bring about constructive changes in local land use decision-making.

Reference for Further Information: NE/EFC Website: http://efc.muskie.usm.maine.edu.

Region 2 Environmental Finance Center at Syracuse University

Description: The Syracuse University Environmental Finance Center (EFC) is located in Region 2 within the Center of Excellence in Environmental and Energy Systems. This EFC provides training, technical assistance, and outreach services to State and local officials related to financing environmental initiatives. Its areas of focus include the promotion of full-cost pricing of environmental services, the exploration of environmental service management options such as privatization, and the facilitation of public input processes in local communities.

The EFC's projects and accomplishments are in areas including asset management, technical assistance, and partnerships. In 2005, it worked with EPA to co-host an asset management workshop in Syracuse. Ongoing initiatives of the EFC include a Public Management and Finance Program designed to offer a more comprehensive and holistic way for communities to benefit from technical assistance, and a partnership with the New York (NY) Department of Environmental Conservation, NY Rural Water Association and NY Water Environment Association to provide a series of wastewater informational sessions for local government representatives.

Reference for Further Information: Syracuse Environmental Finance Center Website: http://www.efc.syracusecoe.org.

Region 3 Environmental Finance Center at the University of Maryland

Description: The University of Maryland's Environmental Finance Center (EFC) is located in Region 3 and is hosted by the University of Maryland's Institute for Governmental Service. The mission of the EFC is to provide communities with the tools and information needed to manage change for a cleaner environment and an enhanced quality of life. This EFC is community-based, with a goal of developing livable communities within the Chesapeake Bay region that promote clean, safe neighborhoods and foster a sense of place for all citizens.

Through strengthening the capacity of local decision-makers to analyze environmental problems and develop innovative and effective ways of financing environmental efforts, the EFC strives to be a leader in community development and watershed protection. This EFC does significant work in the following areas: training for water utility systems managers, working with communities on storm water management, and community financing for local land and water protection.

Reference for Further Information: University of Maryland Environmental Finance Center Website: http://www.efc.umd.edu/who.html.

Region 4 Environmental Finance Center at the University of Louisville

Description: One of two EFCs located in Region 4, the University of Louisville's Environmental Finance Center (EFC) is part of the University's Center for Environmental Policy and Management. The EFC's primary service area is the Southeastern United States. The EFC has two broad mandates stemming from U.S. Environmental Protection Agency goals: 1.) to develop more environmentally and economically sustainable alternatives to uncontrolled and unfocused spatial expansion of human settlements, 2.) to improve the efficiency of environmental infrastructure service delivery.

The services offered by the EFC are divided into four main areas: 1.) Practice Guides, 2.) Research Result Notices, 3.) Financial Planning Support for Water Utilities, and 4.) Brownfields Redevelopment Support. The Practice Guide series is designed for government officials who make or implement policies that influence land use. The Research Result Notices make research results available to public officials via the internet. Through its Financial Planning Support for Water Utilities, the EFC provides training and technical support to operators of water utilities. The EFC's Brownfields Redevelopment Support initiatives are geared towards providing services to facilitate investment in the cleanup and reuse of properties that are perceived to be contaminated.

Reference for Further Information: University of Louisville Environmental Finance Center Website: http://cepm.louisville.edu/org/SEEFC/seefc.htm.

Region 4 Environmental Finance Center at the University of North Carolina

Description: Also located in Region 4, the University of North Carolina Chapel Hill Environmental Finance Center (UNC EFC) is contained within the School of Government. Its primary service area is the southeastern United States and it focuses on service to states and communities. Specifically, the UNC EFC assists communities, provides training and policy analysis services, and disseminates tools and resources on topics such as environmental cost accounting, rate setting, and the development of sustainable cost recovery and institutional management systems.

The UNC EFC reaches local communities through the delivery of interactive applied training programs and technical assistance. The UNC EFC sees one of its major roles as increasing the capacity of other organizations to address the financial aspects of environmental protection. In addition to direct community outreach, the UNC EFC works with decision makers to assess the effectiveness of environmental finance policies at a regional or state level, and to improve those policies as a way of supporting local efforts.

Reference for Further Information: University of North Carolina Environmental Finance Center Website: www.efc.unc.edu/index.html.

Region 5 Great Lakes Environmental Finance Center at Cleveland State University

Description: The Great Lakes Environmental Finance Center (GLEFC) is located in Region 5 at the Maxine Goodman Levin College of Urban Affairs at Cleveland State University. The Great Lakes EFC primarily serves a six-state area, encompassing Ohio, Indiana, Illinois, Michigan, Wisconsin, and Minnesota. The GLEFC acts as a technical assistance, training, and research resource for state and local government, private sector, and non-profit organizations, helping to solve financial problems related to environmental facilities and resources.

One of the Great Lakes EFC's ongoing initiatives is providing Brownfields community site visit advisory services. Brownfields are properties whose financial potential is undermined by real or perceived contamination. The EFC provides professional training to state and local government officials, business executives, and others on environmental finance issues, strategies, and resources, helping clients make more effective use of financial resources for Brownfields redevelopment. The Great Lakes EFC's other projects include financial and economic analyses and market studies to support environmental project planning, development, and implementation, and the publication of research papers and documents.

Reference for Further Information: Region 5 Great Lakes EFC Website: http://www.glefc.org/.

Region 6 Environmental Finance Center at New Mexico Tech

Description: Located in Region 6, the New Mexico Institute of Mining and Technology Environmental Finance Center (NM EFC) is a program of the Institute for Engineering Research and Applications. The primary purpose of the center is to assist state, local, and tribal governments in meeting environmental infrastructure needs and achieving regulatory compliance through state and local capacity building and technical information transfer. Capacity building includes enhancing technical, managerial, and financial capabilities to achieve consistent and sustainable regulatory compliance and to promote and develop sustainable infrastructure.

In particular, the center works to examine alternatives or innovative approaches to meet regulatory compliance and achieve sustainable infrastructure; empower communities to act as the "drivers" and decision-makers for their own projects; present funding alternatives for various types of projects; act as a bridge between federal, state, local, and tribal governments, analyze issues or projects as a neutral entity; gather stakeholder input; and to encourage examination of state or federal programs that inhibit sustainable infrastructure, and offer suggestions of possible approaches that have been used elsewhere. Recently, the NM EFC has been focusing much of its efforts on promoting the concept of asset management to water and wastewater utilities. This program can improve the efficiency and sustainability of utility operations.

Reference for Further Information: New Mexico EFC Website:.
http://nmefc.nmt.edu/home.php

Region 9 Environmental Finance Center at Dominican University

Description: The Dominican University Environmental Finance Center (EFC9) is located within Region 9 and serves Arizona, California, Hawaii, Nevada, Guam, and the Marshall Islands. The mission of EFC9 is three-pronged: 1.) to encourage business to adopt source reduction, pollution prevention and energy efficiency; 2.) to encourage consumers to choose green products and services; and 3.) to help communities promote cleaner business. The EFC carries out many different types of initiatives including organization of conferences and workshops, local economic development, providing hands-on assistance to small businesses, and producing numerous publications.

Specific projects of EFC9 include production of a publication titled <u>Wet Cleaning Guide Booklet</u>, and working with television shows to educate viewers about ways to reduce their impact on the environment. EFC9 also acts as the Region 9 San Francisco Bay Area Green Business Program (GBP) Coordinator. The Bay Area GBP is a partnership of environmental agencies and utilities that assists, recognizes and promotes businesses and government agencies for conserving resources, preventing pollution and minimizing waste. In partnership with the Bay Area GBP, other GBPs located outside the Bay Area, the California Environmental Protection Agency, and the U.S. EPA, EFC9 helps to launch, coordinate, and promote GBPs.

Reference for Further Information: Region 9 EFC Website http://www.efc9.org/.
San Francisco Bay Area Green Business Program (GBP) Website: http://www.greenbiz.ca.gov/.

Region 10 Environmental Finance Center at Boise State University

Description: The Boise State Environmental Finance Center in Region 10 was created in 1995 and is contained within the Department of Public Policy and Administration of the College of Social Sciences and Public Affairs. The Boise State EFC serves the Pacific Northwest and Intermountain states of Alaska, Idaho, Oregon and Washington. The EFC seeks to assist these states and their communities with environmental financing issues, and is committed to helping the regulated community build and improve upon the technical, managerial, and financial capabilities needed to comply with federal and state environmental laws.

The Boise State EFC is an important partner to state and local governments in addressing financing issues related to drinking water and wastewater treatment needs in small communities, helping those communities to meet requirements under the Safe Drinking Water Act of 1996. In addition, the EFC developed a financial planning software program for watersheds called Plan2Fund, and an integrated capital asset inventory and reinvestment analysis software program for water supply systems called CAPFinance. Both of these programs can be downloaded from the EFC's Website and are described on page 5-9 of this section of the Guidebook.

Reference for Further Information: The Environmental Finance Center at Boise State University Website: http://efc.boisestate.edu/efc/.

Boise State University Environmental Finance Center: Plan2Fund

Description: Plan2Fund is a Watershed Planning Tool developed and maintained by the U.S. Environmental Protection Agency's Environmental Finance Center (EFC) at Boise State University. The tool walks users through estimating the costs of their Watershed Program Plan's Goals and Objective, assessing any local matching funds, and determining funding needs to meet their Goals and Objectives. Plan2Fund prompts users to enter specific information on their programs and then generates a series of reports based on that information. The results from Plan2Fund can be used to search for funding sources utilizing the Environmental Finance Center's internet-based Directory of Watershed Resources. The Directory of Watershed Resources is described in this section of the Guidebook.

Reference for Further Information: See
http://efc.boisestate.edu/efc/Tools/Plan2Fund/tabid/104/Default.aspx   For information or assistance, or to request a Plan2Fund CD, call the Boise State EFC toll free at 866-627-9847.

Boise State University Environmental Finance Center: CAPFinance

Description: CAPFinance is an easy-to-use, icon-driven software program that helps public and private water systems with their financial decision-making. It was developed by the U.S. Environmental Protection Agency's Environmental Finance Center (EFC) at Boise State University. The program was developed because small water systems often have trouble estimating and budgeting for future replacement costs. The program helps local officials to understand the impacts of funding capital replacement, and it provides a simple method of analyzing funding options for renewal and replacement of assets. System management in CAPFinance can set the reserve accumulation goal for every component and subcomponent of the water system.

CAPFinance forecasts capital financing needs for 25 years or more. The program helps water utilities inventory capital infrastructure facilities and discover financing requirements, offering unlimited "pay now" or "pay later" scenarios. It produces a report with a detailed view of the future replacement costs and goals for each and every system component. The output from the program can be integrated into financial decision making such as rate setting and capital planning. This planning can help the water supply system to meet customer demands, maintain quality of service, maintain compliance with provisions of Safe Drinking Water Act, and secure the financial resources necessary to fund these efforts.

Reference for Further Information: For information on CAPFinance, or to download a CAPFinance Demo, see the Boise State University EFC's web site at http://efc.boisestate.edu/efc/Tools/AssetManagementwithCapFinance/tabid/90/Default.aspx.

Finance Charrettes

Description: A "finance charrette" is a forum where a regulated entity meets with a panel of finance experts from the public and private sectors, and those experts offer advice and recommendations on finance issues faced by that entity. Adapted by the University of Maryland Environmental Finance Center (and subsequently employed by many of the EFCs) for environmental finance problem solving, the charrette process employs an advisory panel of finance, planning and engineering experts, as well as federal and state officials, who help communities create solutions to their environmental management problems. Environmental Finance Charrettes provide a direct mechanism for ensuring meaningful, constructive two-way communication between higher levels of government and local communities. Typically a charrette lasts a full day beginning with a description of the problems by, for example, officials from a local government. This is followed by question and answer sessions with the panel, and report out by panel members on the actions they recommend as individuals and as a group. The proceedings are taped and the results summarized.

Reference for Further Information: University of Maryland Environmental Finance Center Website: www.efc.umd.edu/charrette.html, E-mail: efc@umd.edu. Also see the description of the University of Maryland Environmental Finance Center in this section of the Guidebook.

Border Environmental Cooperation Commission

Description: The Border Environmental Cooperation Commission (BECC) was created within the context of the North American Free Trade Agreement process and is a sister agency to the North American Development Bank (NADBank). The BECC reviews proposals for environmental projects in the region along the US-Mexico border and certifies them for loan funding by the NADBank (see Section 2.B. of the Guidebook, North American Development Bank). Environmental areas emphasized by the BECC include municipal solid waste management and wastewater treatment. The purpose of the BECC is to help preserve, protect, and enhance the environment of the border region and to achieve sustainable development.

The BECC's operating budget is funded by contributions from Mexico, through the Secretariat of the Environment and Natural Resources, and from the United States, through the Department of State and the Environmental Protection Agency. In addition to its operating budget, the BECC manages the Project Development Assistance Program (PDAP), which receives funding from the United States Environmental Protection Agency. This program allows the BECC to provide border communities with grant funds for water and wastewater projects.

Reference for Further Information: Border Environmental Cooperation Commission (BECC) Website: http://www.cocef.org, E-Mail: becc@cocef.interjuarez.com.
NADBank Website: http://www.nadb.org/.

Rural Community Assistance Corporation

Description: The Rural Community Assistance Corporation (RCAC) is a nonprofit organization dedicated to helping rural communities achieve their goals and visions by providing training, technical assistance, and access to resources. Most RCAC services are provided to low income people and communities with populations fewer than 50,000. Working with governments and community organizations in rural areas, RCAC provides a wide range of development assistance involving housing, environmental services, financial assistance, and information and outreach. RCAC has a loan fund that provides loans to water and wastewater treatment facilities. The five major categories of assistance that RCAC provides to small municipal and nonprofit water systems, wastewater systems and solid waste management programs are: Technical Assistance, Managerial Assistance, Financial Assistance, Networks and Advocacy, Publications, and Training. RCAC's publications are available on its Website.

Reference for Further Information: Rural Community Assistance Corporation Website: http://www.rcac.org/, E-mail: rcacmail@rcac.org, Phone: 916-447-2854.

Self-Help

Description: Self-help is an "in the field" strategy supported by many State government and nongovernmental organizations that helps small communities help themselves in solving their environmental problems. Self-help has proven a highly effective, low-cost approach to providing environmental services and achieving compliance in small communities. It depends heavily on local residents to contribute their time, labor and, on occasion, material and equipment in getting the job done. A local project coordinator or "sparkplug" is essential to success. In the self-help paradigm, State and federal agencies are called upon to move to supporting roles by providing outreach and technical services. The approach offers a proven, viable local alternative to addressing local environmental problems that reduces costs, fits technology to actual needs, builds local capacity, and supports community independence. In many cases, self-help projects can be implemented in a very timely manner due to a decrease in the amount of governmental involvement.

Reference for Further Information: The Self-Help Handbook for Small Town Water and Wastewater Projects, Schautz, Jane W.; and Conway, Christopher M., Rensselaerville Institute, 1995, available through the Rensselaerville Institute, Rensselaerville, NY, Website: http://www.rinstitute.org/shopping/index.php?productID=123, Phone # 518-797-3783.

Circuit Riders

Description: A circuit rider is a dedicated expert who travels on some established regular basis to a number of participating individuals and organizations to provide hands-on technical assistance, professional services, and education. The circuit rider can be either an independent entrepreneur contracting with the participants individually or as a group, or an employee of the participant group acting cooperatively. Furthermore, the circuit rider can work either full or part-time depending on the number of systems participating and the assistance and services provided.

For example, several publicly or privately owned water or other environmental systems may agree to jointly obtain administrative, management, technical, or other services from a common source to meet their common needs. The common source, the circuit rider, addresses the common need such as the collection of samples from each system and delivery of the batch to a lab for testing.

Reference for Further Information: For information on the Ohio T2 Center Circuit Rider Program, see http://www.dot.state.oh.us/LTAP/ltapfaqs.htm, or call them at 614-387-7359, or toll free from locations in Ohio at 877-800-0031. For information on the Massachusetts Department of Environmental Protection Circuit Rider Program, see http://www.mass.gov/dep/water/compliance/cridr.htm, phone # 617-292-5500. To find Circuit Rider Programs in other States, see the Environmental Council of the States (ECOS) Website at http://www.ecos.org/section/states. The Websites of the environmental offices of all U.S. States can be accessed through the ECOS Website.

Cooperatives

Description: A cooperative is an independent association of people and/or groups voluntarily united to meet common needs through a jointly owned and democratically operated venture. For example, several publicly and/or privately owned environmental systems could agree to jointly share administrative, management and technical resources in providing common environmental services. The resulting cost savings are either passed along to users, reinvested in the cooperative venture, or returned to the member systems.

Cooperatives are set up to provide/receive just about any good or service including: business services; financial services; employment service; equipment and farm supplies; insurance; legal and professional services; the marketing of agricultural and other products; and utilities. They are organized in one of three ways: producer-owned, consumer-owned, or worker owned. Cooperatives allow systems to pool not just their resources, but also their technical expertise and knowledge regarding outside sources of assistance.

Reference for Further Information: See the U.S. Department of Agriculture, Rural Development, Business and Cooperative Programs Website at http://www.rurdev.usda.gov/rbs/index.html. Information is also available on the National Cooperatives Business Association Website at http://www.cooperative.org.

Drinking Water State Revolving Fund Capacity Development

Description: The 1996 Amendments to the Safe Drinking Water Act (SDWA) authorize the Drinking Water State Revolving Fund (DWSRF) and spell out requirements for states to prepare capacity development strategies for public drinking water systems. The term "capacity development" refers to a drinking water system's technical, managerial, and financial capabilities to operate now and for the foreseeable future in compliance with all requirements of the SDWA. Systems must demonstrate they have adequate "capacity" or capabilities or that the loan will help them achieve adequate capacity in order to qualify for DWSRF funds.

Each state is required under the SDWA to develop and administer its own capacity development program. The SDWA capacity development provisions are a flexible framework that allows states to design programs that meet the unique needs of their water systems. The programs have components that address newly established water systems, water systems seeking DWSRF loans (as described above) and existing water systems. The range of services available varies from state to state. Interested parties can contact the drinking water regulatory agency in their state to find out about available resources.

Reference for Further Information: The federal capacity development strategy is outlined in Section 1420 of the 1996 SDWA Amendments. Section-by-section summaries, as well as the full text, of the 1996 SDWA Amendments can be found on the U.S. Environmental Protection Agency Website at http://www.epa.gov/safewater/sdwa/laws_statutes.html. For additional information on the SDWA, see http://www.epa.gov/safewater/sdwa/index.html. The full text of the 1996 SDWA Amendments is also available on GPO Access. Go to http://www.gpoaccess.gov/ and search on "Public Law 104-182."

U.S. Environmental Protection Agency:
Water and Wastewater Pricing Website

Description: The U.S. Environmental Protection Agency Water and Wastewater Pricing Website is a valuable source of technical and training information geared towards water systems operators, local utilities, and state and local regulators. It focuses on the "four pillars" of efficient water use and distribution, which are: enhancing utility management, saving water through efficiency measures, cooperative ventures via the watershed approach, and full cost pricing. Full cost pricing plays an important role in providing the public with clean and safe water.

EPA defines full cost pricing as "factoring all costs - past and future, operations, maintenance, and capital costs - into prices." This approach is important to meeting the infrastructure needs of America. The site describes six different types of full cost pricing, and a type of pricing called "lifeline pricing" designed to make water more affordable for low income households. The four types of pricing described on the site as being most effective in encouraging conservation are increasing block rates, time of day pricing, water surcharges, and seasonal rates. The two types described as being less effective in encouraging water conservation are uniform rate structures and flat fee rates.

Reference for Further Information: Water and Wastewater Pricing Website: http://www.epa.gov/water/infrastructure/pricing/index.htm.

U.S. Environmental Protection Agency:
Enforcement Economic Models

Description: The U.S. Environmental Protection Agency (EPA)'s Office of Enforcement and Compliance Assurance develops and maintains enforcement economic models to analyze the financial aspects of environmental enforcement actions. The five models currently available include: ABEL, INDIPAY, MUNIPAY, BEN, and PROJECT. These models evaluate and calculate factors such as costs, economic savings, or ability to afford costs for a variety of regulated entities. ABEL evaluates a corporation's or partnership's ability to afford compliance costs, cleanup costs, and/or civil penalties. INDIPAY and MUNIPAY evaluate the ability of individuals and local governments (or regional utilities), respectively, to afford compliance costs, cleanup costs, and/or civil penalties. ABEL, INDIPAY, and UNIPAY are useful tools to generate information for negotiations. BEN calculates a violator's economic savings from delaying and/or avoiding pollution control expenditures. PROJECT calculates the real cost to a defendant of a proposed supplemental environmental project.

Reference for Further Information: To view information on the models, or to download the models, see the EPA Office of Enforcement and Compliance Assurance Website at http://www.epa.gov/compliance/civil/econmodels/index.html. For installation help, assistance in using the models, or advice in understanding their outputs, call the EPA Enforcement Economic Models Helpline, staffed by an Agency contractor, at 888-ECONSPT or 888-326-6778, or e-mail the Helpline at benabel@indecon.com.

U.S. Environmental Protection Agency:
Green Power Locator

Description: The U.S. Environmental Protection Agency (EPA) Green Power Locator is an online database of information on sources of green power in the United States. The database was created and administered by, the EPA Office of Air and Radiation's Green Power Partnership. EPA defines green power as "electricity that is partially or entirely generated from environmentally preferable renewable energy sources such as solar, wind, geothermal, biomass, biogas, and low impact hydro."

The Green Power Locator is a useful tool for advancing the use of green power. The ways of purchasing green power covered in the database include utility green pricing programs (described in Guidebook Section 8) and renewable energy certificates. As its use becomes more widespread through the use of the types of programs advertised in the database, green power is becoming an increasingly cost effective alternative to electricity generated from fossil fuels.

Reference for Further Information: Green Power Locator Website:
http://www.epa.gov/greenpower/locator/index.htm.

U.S. Department of Energy:
Financial Opportunities Website

Description: The U.S. Department of Energy's Office of Energy Efficiency and Renewable Energy (EERE) offers federal financial assistance to businesses, industries, universities, states, tribes, and others for the development and demonstration of renewable energy and energy efficient technologies. EERE's Financial Opportunities Website helps eligible parties to find and apply for the financial assistance that EERE offers. The Website also provides direct links to current and past solicitations of specific financial awards for businesses, industries, and universities.

The specific types of financial assistance opportunities described on the Website include grants, cooperative agreements, continuation and renewal awards, unsolicited proposals, cooperative research and development agreements, laboratory subcontracts and sub awards. The Website also provides information on several financing measures to help federal energy managers pay for energy related projects, including performance contracts, services contracts, and state and local energy efficiency incentive programs.

Reference for Further Information: See the Financial Opportunities Website at http://www.eere.energy.gov/finacing/. Funding and award process questions can be directed to the Office of Program Execution Support, EERE, ee-3A/Forrestal Building, DOE, Washington, DC 20585, (202) 586-9957 or (202) 586-8180. Also see additional information about specific financial awards at www.grants.gov.

U.S. Department of Energy Alternative Fuels Data Center: State and Federal Incentives and Laws Website

Description: The U.S. Department of Energy, Office of Energy Efficiency and Renewable Energy, Alternative Fuels Data Center is a large online collection of information on alternative fuels and the vehicles that use them. The alternative fuels described in the Data Center are those defined by the Energy Policy Act of 1992. The fuels include biodiesel, electricity (when used to power vehicles), ethanol, hydrogen, methanol, natural gas, and propane.

The "State and Federal Incentives and Laws" Website gives access to the incentives that governments provide to encourage people to reduce oil consumption through the use of alternative fuels and vehicles. It also allows access to the laws and regulations that governments use to ensure that transportation fuels are used in safe and efficient ways. Finally, the Website provides access to points of contact in Federal departments/agencies, the States, the District of Columbia, the Virgin Islands, and Puerto Rico. Contributors to the Website include the U.S. Department of Energy, U.S. Department of Transportation, U.S. Environmental Protection Agency, trade associations, professional societies, auto manufacturers, fuel providers, and universities.

Reference for Further Information: See the Alternative Fuels Data Center's State and Federal Incentives and Laws Website at: http://www.eere.energy.gov/afdc/laws/incen_laws.html.

National Technical Assistance Programs

Description: There are a growing number of national nonprofit technical assistance programs that facilitate the financing and implementation of environmental projects and programs. Such programs can include non-profit organizations ranging from environmental media-based associations to community-focused groups. They can also include university-based groups, professional associations and organizations, and cooperative networks.

Some examples of this type of organization include the American Waterworks Association, the Environmental Protection Agency (EPA) network of nine university-based Environmental Finance Centers (EFCs), the National Rural Water Association, the Rural Community Assistance Programs, and the National Environmental Services Center. Many national technical assistance programs have accumulated considerable experience and developed significant technical expertise in dealing with communities and their environmental and financing problems.

Reference for Further Information: American Water Works Association (AWWA) Website: http://www.awwa.org/, Phone: 303-794-7711. U.S. EPA Environmental Finance Center (EFC) Network Website: www.epa.gov/efinpage, Phone: 202-564-5001. National Rural Water Association (NRWA) Website: http://www.nrwa.org/, Phone: 580-252-0629. Rural Community Assistance Programs (RCAP) Website: http://www.rcap.org/, Phone: 202-408-1273 and 888-321-7227. National Environmental Services Center (NESC) Website: http://www.nesc.wvu.edu/, Phone: (800) 624-8301 and (304) 293-4191. Also see the descriptions of the EFC Network, the NRWA, and the NESC in this section of the Guidebook.

National Rural Water Association

Description: The National Rural Water Association (NRWA) is a nonprofit organization made up of State Rural Water Associations. The NRWA provides support services to its State Associations, who have more than 24,550 water and wastewater systems as members. The utilities that are members of NRWA serve populations of 10,000 or less, which represents 94% of all water systems in America.

NRWA's State Rural Water Associations offer a variety of state specific programs, services, and member benefits. Additionally, each State Association provides training programs and on-site assistance in the areas of operation, maintenance, finance, and governance to water and wastewater system personnel. Also, the NRWA operates the International Rural Water Association (IRWA). The IRWA's mission is to help improve water quality and, in turn, public health, in developing countries. The IRWA's primary goal is to make available the economic distribution of water treatment, training, and technical assistance to people in rural communities.

Reference for Further Information: National Rural Water Association Website: http://www.nrwa.org/, E-mail: info@nrwa.org, Phone: 580-252-0629.

Rocky Mountain Institute

Description: The Rocky Mountain Institute (RMI) is an entrepreneurial nonprofit organization that fosters the efficient and restorative use of resources. RMI works with businesses, civil society, and governments to design integrative solutions that help create long-term prosperity. It shows businesses, communities, individuals, and governments ways to create wealth and employment, protect and enhance natural and human capital, increase profit and competitive advantage, and enjoy many other benefits, largely by increasing the efficiency of their processes. RMI's work is independent and non-adversarial, with a strong emphasis on market-based solutions.

Since 1982, RMI has worked with corporations, governments, communities, and citizens to increase their resource efficiency, in turn increasing productivity and profits. Some of the environmental areas in which RMI has applied its efficiency expertise include household and commercial/industrial energy utilization; clean energy and climate protection; green development; public and private water use; water infrastructure and system planning, benefits, and costs; and stream restoration. RMI has provided a range of consulting and research advisory services in these and other areas. Senior RMI staff has addressed audiences at events and institutions such as the World Economic Forum, the World Bank, and the National Academy of Sciences. Senior staff has also provided private briefings or expert testimony to heads of state, corporate boards, utility commissions, top military leaders and staff colleges, elite business and law schools, and governmental advisory boards.

Reference for Further Information: Rocky Mountain Institute Website: http://www.rmi.org/.

U.S. Environmental Protection Agency:
Water Efficiency Market Enhancement Program

Description: The U.S. Environmental Protection Agency (EPA) Water Efficiency Market Enhancement Program works to promote the use of more water-efficient products and practices in businesses and homes across the country. The Market Enhancement Program does this by reaching out to various organizations and fostering public-private partnerships. The Program seeks to help consumers and commercial/institutional buyers differentiate among products in the marketplace, helping them to buy the most water efficient products. In doing this, it strives to reduce water demand and realize major environmental, public health, and economic benefits by helping to improve water quality, maintaining aquatic ecosystems, and protecting drinking water resources. While this is a new EPA program, the Agency hopes to make water-efficient products and systems the preferred choice among consumers and buyers. The Market Enhancement Program has a list of products being evaluated for inclusion in the Program on its Website.

Reference for Further Information: Water Efficiency Market Enhancement Program Website http://www.epa.gov/owm/water-efficiency/products_program.htm, Phone 202-564-0637, Fax 202-501-2396, E-mail address water_efficiency@epa.gov.

Environmental Council of the States

Description: The Environmental Council of the States (ECOS) is a national nonprofit, nonpartisan association of state and territorial environmental organization leaders. The objective of ECOS is to improve the capability of state environmental organizations and their leaders to protect and improve human health and the environment. The Council defines its role as: 1) articulate, advocate, preserve, and champion the role of the states in environmental management; 2) provide for the exchange of ideas, views, and experiences among states and with others; 3) foster cooperation and coordination in environmental management, and 4) articulate state positions to Congress, federal agencies, and the public on environmental issues.

Other than the general membership, ECOS is run by a 28 member executive committee led by four officers: a president, vice president, secretary/treasurer, and past president. ECOS has several finance related work groups: the ECOS-DOD Sustainability Work Group, the Energy Efficiency Subcommittee, the Funding Gap Work Group, and the Long Term Stewardship Work Group. ECOS's Website provides an information portal into every state environmental department/organization and thus into their financing operations. The Website is an information source on "financial tools" such as grants, loans, and environmental programs offered and administered on the state level.

Reference for Further Information: Environmental Council of the States Website: www.ecos.org. The Website has a "contact us" form. ECOS phone #: 202-624-3660.

Pollution Prevention Resource Exchange

Description: The Pollution Prevention Resource Exchange (P2Rx) is a national consortium of eight regional centers that provide states, local governments, and technical service providers with pollution prevention (P2) information, services, and networking opportunities. P2Rx is funded in part through grants from the U.S. Environmental Protection Agency. For P2Rx purposes, P2 means "source reduction," as defined under the *Pollution Prevention Act of 1990*. Source reduction includes practices that reduce or eliminate pollution via improved efficiency in the use of raw materials, energy, water, or other resources. P2Rx disseminates information on topics including technologies, publications, green home design and construction, and tribal pollution prevention. Services P2Rx provides also include research assistance, and the maintenance of clearinghouse of P2 related requests for funding proposals available on its Website. The P2Rx Website also has a database of mercury reduction programs, a directory of P2 programs, and links to the eight regional centers.

Reference for Further Information: Pollution Prevention Resource Exchange Website: http://www.p2rx.org/, Phone: 402-552-6259. The full text of the *Pollution Prevention Act of 1990* is available via the U.S. Environmental Protection Agency Website at http://www.epa.gov/opptintr/p2home/p2policy/act1990.htm.

Energy Efficiency Organizations

Description: Energy Efficiency Organizations (EEOs) are non-governmental, non-profit organizations that provide energy efficiency-related information, education, outreach, and technical assistance. In some cases, they actively promote improved energy efficiency and the use of energy efficient practices, products and services. EEOs may be international, national, regional, or state-based in scope. There are many different types of organizations that can be classified as an EEO ranging from consortiums or alliances comprised of utilities, research organizations, and state energy offices to national housing coalitions, environmental organizations, and regional energy programs.

A significant amount of the work of EEOs directly impacts the environmental arena especially with regards to air pollution, climate change issues, and sustainable development. Some of the many possible examples of EEOs include the World Energy Efficiency Association, Alliance to Save Energy, American Council for an Energy Efficient Economy, Consortium for Energy Efficiency, Green Building Council, Sierra Club (energy), Northwest Energy Coalition, Colorado Energy, Sustainable Minnesota, and New Buildings Institute, Inc.

Reference for Further Information: Websites with links to lists of EEOs include http://www.naima.org/pages/resources/links2.html, http://eetd.lbl.gov/eXroads/ngo.html, http://www.ecoiq.com/onlineresources/center/energy/efficiency/nonprofit.html. and www.eere.gov/EE/buildings-trade.html. A list of EEOs can also be acquired by going to www.envirolink.org and searching on "energy efficiency organizations."

U.S. Department of Energy: Industrial Assessment Centers

Description: The Industrial Assessment Centers are a part of the U.S. Department of Energy Office of Energy Efficiency and Renewable Energy's Industrial Technologies Program (ITP), a program that leads national efforts to improve industrial energy efficiency, productivity, and environmental performance. The Centers provide eligible small and medium-sized manufacturers with comprehensive energy, waste, and productivity assessments free of charge. They then provide detailed recommendations to the manufacturers, helping them to identify opportunities to improve productivity, reduce waste, and save energy. Manufacturers have benefited from cost savings in excess of tens of thousands of dollars through implementing these recommendations.

The Centers are located at 26 universities throughout the Unites States. Teams of engineering faculty work with upper class undergraduate and graduate engineering students to perform the assessments. An advantage in using students to do this work is that they receive unique hands-on assessment training and an increased knowledge of industrial process systems, plant systems, and energy systems, making them effective energy professionals who are highly attractive to employers. With over 2,300 graduates, the Industrial Assessment Centers constitute a resource for corporate recruiters seeking experienced energy engineers.

Reference for Further Information: See the Industrial Assessment Centers Website at http://www.eere.energy.gov/industry/bestpractices/iacs.html and the student/alumni Website at www.iacforum.org.

Catalog of Federal Domestic Assistance

Description: The Catalog of Federal Domestic Assistance (CFDA) is an online database listing all federal programs, projects, services, and other initiatives providing financial benefits and other forms of assistance to the public. The CFDA contains detailed information on financial and non-financial assistance programs administered by departments, agencies, commissions, and other federal government establishments. The CFDA includes grants and loans that can provide financial assistance. The document also includes forms of non-financial assistance, such as loans of equipment and provision of specialized services. The types of information provided for each assistance service includes program objectives, types of assistance, use and restrictions, eligibility requirements, application and award processes, and post assistance requirements. In addition, financial information such as range and average of financial assistance is provided, followed by program accomplishments; regulations, guidelines, and literature; information contacts, and examples of funded projects.

Reference for Further Information: Catalog of Federal Domestic Assistance Website: http://12.46.245.173/cfda/cfda.html. A printed copy of the Catalog may be purchased from the Government Printing Office (GPO) by calling toll free 1-866-512-1800 or by logging on to the GPO's website at http://bookstore.gpo.gov/.

Database of State Incentives for Renewable Energy

Description: The Database of State Incentives for Renewable Energy (DSIRE), established in 1995, is a comprehensive electronic source of information on state, territorial, local, utility, and selected federal incentives that promote the use of renewable energy technologies. The renewable energy technologies it promotes include solar (active and passive); wind power; biomass; cogeneration, or combined heat & power (CHP); fuel cells; geothermal; hydroelectric; ocean thermal; and photovoltaics.

DSIRE is an ongoing project of the Interstate Renewable Energy Council (IREC), funded by the U.S. Department of Energy and managed by the North Carolina Solar Center. The Interstate Renewable Energy Council is a non-profit organization whose mission is to accelerate the use of renewable energy sources and technologies. The North Carolina Solar Center is a program of North Carolina State University's College of Engineering's Industrial Extension Service. It works to advance the use of renewable energy resources to ensure a sustainable economy.

Reference for Further Information: Database of State Incentives for Renewable Energy Website: http://www.dsireusa.org/. Contact information for the Interstate Renewable Energy Council, the U.S. Department of Energy, and North Carolina State University's Solar Center can be found on the DSIRE Website.

U.S. Environmental Protection Agency: Catalog of Federal Funding Sources for Watershed Protection

Description: The Catalog of Federal Funding Sources for Watershed Protection is a searchable online database of funding sources available for a range of different watershed protection activities. The Catalog contains information on more than 80 federal funding sources covering a wide variety of grant, loan and cost-sharing assistance programs. The Catalog's Website also has a link to "other funding sources" that provides users with an extensive listing of public and private sector sources, including publications and funding-related web sites, that could help secure additional sources of funding. The Catalog was created, and is administered, by the U.S. Environmental Protection Agency (EPA) Office of Wetlands, Oceans, and Watersheds.

On the Website, the user has a choice of two different types of searches. One type of search is based on subject matter criteria, and the other is based on words in the title of the funding program. Criteria searches consist of the type of organization, type of assistance sought, and keywords. The information provided for each program in the Catalog includes contact information, funding history, typical past awards, eligibility requirements, application deadlines, and matching funds/criteria requirements.

Reference for Further Information: See the Catalog's Website at: http://www.epa.gov/watershedfunding, or alternatively, at http://cfpub.epa.gov/fedfund/.

Boise State University Environmental Finance Center:
Directory of Watershed Resources

Description: The Directory of Watershed Resources is a searchable online database for sources of watershed restoration funding. It produced and maintained by the U.S. Environmental Protection Agency (EPA) Environmental Finance Center (EFC) at Boise State University. The Directory includes information on funding programs from Federal, State (Oregon, Washington, Idaho, Alaska, Connecticut, Maine, Massachusetts, New Hampshire, Rhode Island, and Vermont), private, and other sources. Users can query the information in a variety of ways, including agency source or keyword, or they can opt to do a more detailed search.

The Boise State EFC points out on its Website that this Directory is a work-in-progress. Information is added to it and updated regularly. The EFC strives to maintain the most current information, but it still recommends that users visit the funding program Websites or contact the funding program administrators for the most up-to-date information. The Environmental Finance Center at the University of North Carolina has also developed a similar database that includes funding information for Alabama, Florida, Georgia, Mississippi, North Carolina and South Carolina.

Reference for Further Information: See http://sspa.boisestate.edu/efc/ and click on "Directory of Watershed Resources: Search online for Funding Sources" under "What's New," or go directly to the Directory of Watershed Resources at: http://efc.boisestate.edu/. Contact: Crystal Morehead at the Boise State EFC, Phone: (208)426-1567, E-mail: cmorehea@boisestate.edu. There is a description of the Boise State EFC in this section of the Guidebook.

Although federal government programs and other national efforts are extremely important funding sources, they can not possibly address all of the current need. Furthermore, these funding sources are decreasing over time. With the decrease in these funding sources, communities have to seek other methods of meeting their critical environmental needs, such as: seeking state funding or leveraging their limited resources through innovative local efforts, including new volunteer programs. There are a wide variety of state programs available to local communities, nonprofits, and other groups to assist them in reaching their environmental goals. These programs are becoming increasingly more responsive to the needs of local communities and the states are using a wide variety of innovative measures to assist in attaining environmental finance goals.

This section of the Guidebook provides communities with information regarding the financial resources that are available to them at the state level. In an effort to achieve a balanced geographical representation, the Guidebook includes at least one state financial tool from each of the U.S. Environmental Protection Agency's 10 regions. Often, these tools are representative of the tools available in other states. Specific information on the tools in other states can be found by contacting the environmental regulatory agency in your state or by looking at the agency's web page. The website for every U.S. State environmental office can be accessed through the Environmental Council of the States website at http://www.ecos.org/section/states. Additional information on state financial tools can be accessed through the Database of State Incentives for Renewable Energy at http://www.dsireusa.org/.

The state programs showcased in this section provide a representation of what is available at the state level. Every effort was made to capture creative and innovative programs, but there was no intent to indicate that these are the states' "best" programs. By highlighting some of the more creative programs, we hope this section of the Guidebook can not only serve local communities seeking help, but also assist state governments in researching potential new solutions or approaches to solving environmental problems. Additionally, there are many creative ideas and programs portrayed that could potentially be incorporated at the federal level.

The growing pressure created by reduced federal funding has stimulated creativity and innovation on the part of states. State governments are very close to the environmental problems they are required to address and are exceptionally aware of what does and does not work. State and local governments are increasingly using their own funds, or a combination of their own funds in union with other funding sources, to get their projects initiated.

The programs presented in this section are primarily funded through state resources - not federal funds. However, in some cases, state programs are presented that are funded and implemented through a myriad of sources - state funds, foundations, federal funds, tax incentives, volunteer action, etc. Many states are providing effective "integrated" programs that offer communities several forms of assistance through one single state office - loans, grants, circuit-riders, assistance from professionals, publications, etc. If one form of assistance does not help the community, another probably will.

Many state offices work with applicants from the beginning of their projects to the end, helping them to find financial assistance even if they do not qualify for the programs offered through the state. Some states have chosen to work in partnership with their communities in a partnership until the environmental problem is solved or the project is off the ground, completed, and self-sustainable.

1.   Co-Funding
2.   Solar 4R Schools Program
3.   Renewable Energy Credits
4.   State Conservation Tax Credits
5.   State Energy Efficiency Tax Incentives
6.   Unit-Based Pricing for Solid Waste Collection
7.   Utility Rebate Programs
8.   State of California: Electronic Waste Recycling Fee
9.   State of Colorado: Easement Program
10. State of Maryland: Bay Restoration Fund
11. State of Minnesota: Capital Assistance Program
12. State of Montana: Renewable Resource Grant and Loan Program
13. State of Nebraska: Illegal Dumpsite Cleanup Program
14. State of Nevada: Financial Assistance for Drinking Water Systems Program
15. State of New Mexico: Clean Energy Grants Program
16. State of New York: Green Building Tax Credit
17. State of North Carolina: Clean Water Management Trust Fund
18. State of Ohio: Water Pollution Control Loan Fund
19. State of Oregon: Truck Engine Tax Credit
20. State of Pennsylvania: Growing Greener Program
21. State of Rhode Island: Aqua Fund
22. Net Metering

Co-Funding

Description: State and local governments often use co-funding, which is the combining of many different forms of funding, to finance environmental protection initiatives. An example of co-funding is the combining of federal and state loans, and perhaps grants as well, to fund the same project. Co-funding opportunities are particularly applicable and advantageous to small communities for funding wastewater and drinking water treatment, nonpoint source pollution prevention, and other environmental protection initiatives. The potential use of co-funding for environmental projects is great, especially if an agency is willing to take the lead in coordinating different funding sources, cycles and procedures. The New York State Water and Sewer Infrastructure Co-Funding Initiative is an example of one of the many state programs helping communities to find sources of government funding for water and sewer projects.

Reference for Further Information: Contact state and local government offices with inquiries-the Website for the environmental office of every state in the U.S. can be accesses through the Environmental Council of the States Website at http://www.ecos.org/section/states. New York State Water and Sewer Infrastructure Co-Funding Initiative Website: http://www.nycofunding.org/newcofund/.

Net Metering

Description: Net metering creates a financial incentive for utility customers to invest in power generation utilizing renewable resources. Net metering allows homeowners to receive the full retail value of the electricity that their renewable energy systems generate. Using net metering, homeowners with renewable energy systems, such as windmills and solar photovoltaics, can offset their electric bills with any excess electricity they produce. However, net metering is only available in states with net metering laws. In states without net metering laws, utilities are still required under federal law (18 CFR Part 292) to purchase excess electricity generated by their customers, however the purchase would be at the wholesale price, which is significantly lower than the retail price.

Reference for Further Information: American Wind Energy Association Website: http://www.awea.org/faq/netbdef.html. U.S. Department of Energy Website: http://www.eere.energy.gov/solar/net_metering.html and http://www.eere.energy.gov/greenpower/markets/netmetering.shtml. Database of State Incentives for Renewable Energy (DSIRE) Website: http://www.dsireusa.org/.

Solar 4R Schools Program

Description: The Solar 4R Schools Program is an initiative of the Bonneville Environmental Foundation, providing grants for the installation of 1.1 kW demonstration solar and data acquisition systems in middle and high schools in the Pacific Northwest. The Program is intended specifically for schools interested in increasing the visibility of renewable energy sources. Successful projects pursuant to the Program's goals offer aggressive education and outreach plans in an attempt to overcome barriers to widespread adoption of photovoltaics. Grants are awarded year-round in accordance with project needs and schedules. Any middle or high school located in Oregon, Washington, Idaho, or Montana may submit a Letter of Enquiry to the Bonneville Environmental Foundation describing a proposed renewable energy project and requesting funding for it.

Reference for Further Information: Bonneville Environmental Foundation Website: http://www.b-e-f.org/grants/solar.shtm. The Proposal Guidelines and Letter of Enquiry form are available for downloading on the Website. Letters of Enquiry should be e-mailed to info@b-e-f.org. All other correspondence should be mailed to: Renewable Energy Programs, Bonneville Environmental Foundation, 133 SW 2nd Ave., Suite 410; Portland, OR; 97204.

Renewable Energy Credits

Description: Renewable Energy Credits, or "Credits," are tradable certificates of proof, in which each Credit is a verification that one kilowatt-hour (kWh) of electricity has been generated by a renewable energy source. Credits are a separate commodity from the power itself. The Credits are used to fulfill requirements under state Renewables Portfolio Standards (RPS's). Each state with an RPS has a requirement that a minimum percentage of renewable energy be included in the portfolio of electricity sources distributed to customers within state boundaries. That percentage is often increased each year. For the purposes of RPS's, what qualifies as renewable energy varies from state to state, but generally includes solar thermal electric, photovoltaics, wind, biomass, hydroelectric, geothermal electric, tidal energy, wave energy, ocean thermal, and fuel cells. The renewable energy markets created by the use of Renewables Portfolio Standards and Credits make possible long-term contracts and financing for industries generating power from renewable energy sources.

Reference for Further Information: Wind Energy and Energy Policy Website: http://www.awea.org/policy/rpsbrief.html. For information on the RPS's of specific states, see http://www.dsireusa.org/ and look for "Renewables Portfolio Standard" under "Rules, Regulations, and Policy."

State Conservation Tax Credits

Description: Through state conservation tax credits, landowners are allowed a deduction on their income taxes if they donate land to public or nonprofit entities for conservation. A growing number of states throughout the U.S. offer these tax credits. These credits are frequently used for individual and/or corporate income taxes. The types of donations eligible for state conservation tax credits include easements, fee interests, and water rights transfers. These land donations are a very effective way of protecting open space, particularly if the land becomes state conservation land, which is often protected from eminent domain purchases. The extent of the conservation benefits from each land donation is, however, limited by how well the new landowner acts as a steward for it. For this reason, the person donating the land will often establish a written agreement with the new landowner preventing the land from being sold again and specifying how it can and cannot be used.

Reference for Further Information: State Environmental Resource Center Website: http://www.serconline.org/conservationTaxIncentives/stateactivity.html, Phone: 608-252-9800, E-mail: info@serconline.org. For definitions of easements and fee interests see http://teachmefinance.com, click on "financial terms," and use the search feature.

State Energy Efficiency Tax Incentives

Description: Many U.S. states offer tax incentives to encourage the use of energy efficient appliances, or to promote energy efficiency improvements and renewable energy use in homes and businesses. These tax incentives include income tax deductions and subtractions, sales tax exemptions, and property tax exemptions. Many states offering tax incentives for the purchase of energy efficient appliances require that the appliances be certified under the U.S. Environmental Protection Agency Energy Star Program.

To obtain tax credits for energy efficiency improvements in homes and commercial buildings, states' often require ratings to be obtained from third party agencies such as: Home Energy Rating Systems (HERS), which is used for homes, or Leadership in Energy and Environmental Design (LEED), which is used for commercial buildings. A HERS rating is an assessment of the energy efficiency of a home as compared to a computer-simulated reference house. LEED provides a complete framework for assessing building performance and meeting sustainability goals including energy efficiency and water savings.

Reference for Further Information: State Environmental Resource Center Website: http://www.serconline.org/energytaxincentives.html. Database of State Incentives for Renewable Energy Website: http://www.dsireusa.org/. Energy Star Program Website: http://www.energystar.gov/, information about HERS ratings under "New Homes." The U.S. Green Building Council Website, http://www.usgbc.org/, has information about LEED ratings.

Unit-Based Pricing for Solid Waste Collection

Description: Unit-based pricing is a system in which residents pay for municipal solid waste management services per unit of waste collected rather than through a fixed fee or property taxes. It is also known as variable rate pricing, user pay, or pay-as-you-throw. Communities with unit-based pricing programs report great increases in recycling and reductions in waste, due primarily to the waste reduction incentive created by the pricing structure. Examples of different types of unit-based pricing for solid waste collection include:
1.) Pre-paid bag: Households purchase official, standard-sized trash bags. Only garbage in these official marked bags is collected.
2.) Pre-paid tag/sticker: Households purchase official tags or stickers. The charge covers a specific size, or sizes of, containers. Only garbage containers marked with the tags or stickers are collected.
3.) Weight-based system: Each household pays a set fee per pound of garbage collected.

Reference for Further Information: See the U.S. Environmental Protection Agency (EPA) Office of Solid Waste and Emergency Response (OSWER) Website at: http://www.epa.gov/epaoswer/non-hw/payt/intro.htm. Also see the "contact us" section of OSWER's Website at: http://www.epa.gov/epaoswer/osw/comments.htm.

Utility Rebate Programs

Description: In many locations throughout the United States, local electric utilities offer rebate programs for the purchase and installation of products that reduce the energy consumption of commercial, industrial, and residential users. One commonly used method of funding these types of rebates is to apply an "energy conservation charge" to the bills of utility customers. These energy conservation charges are generally non-taxable. One common usage of these utility rebates is for the purchase of solar photovoltaic systems, although utilities often offer rebates for the purchase of solar water heaters, energy efficient appliances or air conditioners, wind energy systems, various retrofits to improve the energy efficiency of buildings, and solar thermal and geothermal heat pump systems.

Reference for Further Information: See the Database of State Incentives for Renewable Energy (DSIRE) Website at http://www.dsireusa.org/ and click on individual states for more information on utility rebate programs offered throughout the U.S. and in the U.S. territories.

State of California: Electronic Waste Recycling Fee

Description: The California Electronic Waste Recycling Fee law requires retailers to collect a fee when they sell certain video display devices to consumers. It is authorized under California's Electronic Waste Recycling Act of 2003, which was signed into law on September 24, 2003, and amended by SB 50 on September 29, 2004. The fee was established to fund a program for consumers and the public to return, recycle, and ensure the safe and environmentally sound disposal of video display devices, reducing the amount of hazardous waste going into landfills. The fee must be collected by the retailer for each Covered Electronic Device (CED) sold. The fee can be collected at the store, by mail order, or over the Internet. CEDs include computer monitors and televisions and the fee ranges from $6.00 to $10.00 per CED. Retailers are permitted to retain three percent of the fees they collect.

Reference for Further Information: See the State of California Website at http://www.boe.ca.gov/sptaxprog/ewfaqsgen.htm. Contact: The California Board of Equalization at 800-400-7115, or via an online form at http://www.boe.ca.gov/info/contact.htm. California's Electronic Waste Recycling Act of 2003 is in Stats. 2003, chapter 526 - SB 20, and it is amended by SB 50 (Stats. 2004, chapter 863).

State of Colorado: Conservation Easement Tax Credit Program

Description: Established in 2001, the Colorado Conservation Easement Tax Credit Program allows landowners who donate conservation easements on their land to receive credits on their Colorado state income taxes. Any landowner who cannot use the credit (i.e., does not have a high enough taxable income) can sell the credit, at a reduced rate, to someone else who can use it. The Colorado Conservation Trust (CCT) recruits buyers of the credits, generates positive press coverage for the program, maintains quality control, and works with credit buyers to direct their tax savings towards conservation projects.

To receive the tax credit, a landowner must place a conservation easement on his/her property which permanently restricts development on the land. By placing an easement on the property, the landowner receives a state tax credit of up to $260,000. The first $100,000 is a dollar-for-dollar value; and the rest of the credit is a 40 cents on the dollar value. Typically the seller receives 80% of the value of the credit and the buyer pays 85-90% of the value, with the remainder going to the broker and to conservation organizations.

Reference for Further Information: Colorado Conservation Trust (CCT) Website: http://www.coloradoconservationtrust.org/cctprograms/initiatives_taxcredit.php, Phone: 720-565-8289, E-mail: coctinfo@coct.org. To access land conservation finance related articles including "Changes to Colorado's Conservation Income Tax credit Law" by Jay, Jessica E., see: http://www.privatelandownernetwork.org/library/ and look under "Related Resources."

State of Maryland: Bay Restoration Fund

Description: On May 26, 2004, the State of Maryland Senate Bill 320 (Bay Restoration Fund) was signed into law. The bill created a dedicated fund, which is financed through a $2.50 monthly fee that is levied upon all users of centralized wastewater treatment facilities., The fund is being used to upgrade Maryland's centralized wastewater treatment plants to include enhanced nutrient removal (ENR) technology to achieve a wastewater effluent quality of 3 mg/l total nitrogen and 0.3 mg/l total phosphorus. A similar fee levied upon septic system users is being used to upgrade onsite septic systems and plant cover crops to reduce the nitrogen loading to the Chesapeake Bay. This Fund is administered by the Maryland Department of the Environment and includes the establishment of an advisory committee. Fees from users of centralized wastewater treatment facilities generate an estimated $65 million per year, while an additional $12.6 million is collected from homes using septic systems.

Reference for Further Information: Bay Restoration Fund Website:
http://www.mde.state.md.us/Water/CBWRF/index.asp.

State of Minnesota: Capital Assistance Program

Description: Through the Capital Assistance Program (CAP), the Minnesota Office of Environmental Assistance (OEA) provides grants to fund the capital costs of building solid waste processing and resource recovery facilities. Many other related projects are also funded. The CAP is open to local government agencies, and solid waste management and sanitary districts. The OEA encourages projects that involve public-private cooperation.

Eligible projects for CAP grants include solid waste processing facilities, waste-to-energy facilities, compost facilities for yard waste and organics, recycling and resource recovery facilities, and household hazardous waste collection and disposal facilities. More than 90 CAP grants have been awarded since 1985. During the 2005 Legislative Session, $4 million in bond funds were appropriated for CAP funding.

Reference for Further Information: Capital Assistance Program (CAP) Website:
http://www.moea.state.mn.us/grants/cap.cfmhttp://www.fs.fed.us/spf/coop. For application forms and CAP Procedures Manuals call Mary James at 651-215-0194. The Office of Environmental Assistance can also be reached at 1-800-657-3843.

State of Montana: Renewable Resource Grant and Loan Program

Description: The State of Montana's Renewable Resource Grant and Loan Program provides funding for the conservation, management, development and preservation of Montana's natural resources. Administered by the Montana Department of Natural Resources and Conservation (DNRC), the program provides funding to government entities, individuals, and groups for a variety of projects including groundwater studies, irrigation rehabilitation, water and soil conservation, municipal drinking water improvements, wastewater treatment, forest enhancement, and renewable energy projects.

Financial awards issued through this program include: Renewable Resource Grants, Emergency Grants, Project Planning Grants, loans, and private grants and loans. Renewable Resource Grants awarded to state and local governments are capped at $100,000. Project Planning Grants are limited to $10,000. Emergency grants are awarded to government entities and capped at $30,000. Private grants and loans are awarded to individuals and groups and are capped at $5,000. Loans are limited by the applicant's ability to repay within 20 years. Grant applications are due May 15 of even numbered years. Loan applications are accepted year-round.

Reference for Further Information: Montana DNRC Website: http://www.dnrc.mt.gov/cardd/ResDevBureau/renewable_grant_program.asp. Renewable Resource Grant and Loan Program contact information: Phone: 406-444-6839, E-mail: Bob Fischer at rfischer@mt.gov, o r Pam Smith at pamsmith@mt.gov.

State of Nebraska: Illegal Dumpsite Cleanup Program

Description: Nebraska's Illegal Dumpsite Cleanup Program was established in 1997 to fund the cleanup of illegal non-hazardous waste sites. It is administered by the Nebraska Department of Environmental Quality (DEQ). The program reimburses political subdivisions for the cleanup of solid waste that is dumped along public roadways or ditches. It also provides funding for the cleanup of waste that spills from these illegal roadside dumps to contiguous private property. Through the program, items such as household waste, white goods, construction and demolition waste, and furniture are removed from the illegal sites and disposed of in permitted facilities or recycled. To finance this program, the Nebraska DEQ allocates five percent of the revenue from the tipping fee that is collected for waste disposal at licensed landfills statewide. The fund generates approximately $125,000 per year.

Reference for Further Information: For more information on Nebraska's Illegal Dumpsite Cleanup Program and a similar program in Kansas, see the U.S. Environmental Protection Agency Website at http://www.epa.gov/region07/waste/solidwaste/illegal_dumping.htm. For additional information on the Nebraska program, see http://www.deq.state.ne.us/, and go to Ch. 5, p. 40 of the 2005 Annual Report to the Legislature. Nebraska DEQ Phone: 402-471-2186, E-mail: MoreInfo@NDEQ.state.NE.US. For additional information on the Kansas program, see: http://www.kdheks.gov/waste/illegal_dump_cleanup/Illegal_dump_letter.pdf.

State of Nevada: Financial Assistance for Drinking Water Systems Program

Description: The Nevada Financial Assistance for Drinking Water Systems Program, also called the "Assembly Bill (AB) 198 Grant Program," provides grants to purveyors of water systems. The AB 198 Grant Program is offered through the Bureau of Water Pollution Control, which is part of the State of Nevada Division of Environmental Protection. The Program is designed to help communities bring their water systems into compliance with Nevada State Health Board and Safe Drinking Water Act regulations.

Grants are made to water districts, counties, and incorporated towns. Grant funds are used to pay for capital improvements to publicly-owned community water systems and publicly-owned nontransient water systems. These grants are leveraged; grant awards are required to cover no less than 57% and no more than 87% of the eligible costs for the project. The remaining project expenses are the responsibility of the grantee.

Reference for Further Information: Nevada Division of Environmental Protection Website: http://www.ndep.nv.gov/bffwp/grants01.htm. There is a staff directory on the Website.

State of New Mexico: Clean Energy Grants Program

Description: New Mexico's Clean Energy Grants Program, established in 2004, supports the development of renewable energy, energy efficiency, and alternative transportation fuel technologies. It is administered by the Energy Conservation and Management Division of the New Mexico Energy, Minerals and Natural Resources Department (ECMD-EMNRD). Grants are awarded to municipalities, county governments, state agencies, public schools (K-12), post-secondary educational institutions, and tribal entities throughout the State of New Mexico.

Capital projects funded by these grants are required to meet performance measures established for the program, including a 5% reduction in energy consumption in building projects or a 15% increase in alternative fuel usage. Educational and non-capital projects must provide one of the following benefits: a) increasing the demand for clean energy, or b) advancing commercialization and widespread application of clean energy technologies. The maximum amount of any grant awarded under this program is $200,000.

Reference for Further Information: New Mexico ECMD-EMNRD Website: http://www.emnrd.state.nm.us/emnrd/ecmd/CleanEnergyBills2005/CleanEnergyGrantsProgram2005.htm. Contact: Louise Martinez at the ECMD-EMNRD at 505-476-3310. Information on this grant program is also available on the Database of State Incentives for Renewable Energy (DSIRE) Website at http://www.dsireusa.org/. There are also many renewable energy and energy conservation grant programs offered in different U.S. states listed on the DSIRE Website.

State of New York: Green Building Tax Credit

Description: The New York Green Building Tax Credit program was signed into law in 2000 to provide an incentive encouraging the construction of new buildings designed in a way that conserves energy and minimizes environmental impact. Green Building tax credits are offered to building owners and tenants who invest in technologies and construction practices that meet requirements spelled out in the Green Buildings Tax Credit program regulations (6NYCRR Part 638). These "green building" practices include water efficient landscaping, indoor pollutant source control measures, and the use of recycled building materials and energy efficient technologies. Owners and tenants of buildings such as hotels, office buildings, and residential multi-family buildings are eligible to apply for these tax credits.

Reference for Further Information: New York State Dept. of Environmental Conservation Green Building Initiative Website: http://www.dec.state.ny.us/website/ppu/grnbldg/#summary. Natural Resources Defense Council Website: http://www.nrdc.org/cities/building/nnytax.asp. Information on Green Building Tax Credits in other states is available on the State Environmental Resource Center Website: http://www.serconline.org/grBldg/stateactivity.html.

State of North Carolina: Clean Water Management Trust Fund

Description: In 1996, the North Carolina General Assembly established the Clean Water Management Trust Fund (CWMTF) to help finance projects that specifically address three principal objectives: 1) the restoration of degraded waters, 2) the protection of unpolluted waters, and 3) the establishment of riparian buffers. The CWMTF is a voluntary, incentive-based water quality program. Eligible applicants for CWMTF grants are: 1) state agencies, 2) local governments or other political subdivisions of the state, or a combination of such entities; and 3) conservation nonprofits. Priority for grants is given to economically distressed units of local governments. The CWMTF funds about one third of the requests it receives, with a total of $535.4 million in grant funds awarded between 1996 and 2005. No matching share is required, although the fund encourages partnerships with other programs working to protect waterways or adjoining lands. Through partnerships, the CWMTF funds have leveraged over $810.3 million in additional funds.

Reference for Further Information: The CWMTF is authorized under Article 18, Chapter 113A of the North Carolina General Statutes. For details and application information see the CWMTF Website at http://www.cwmtf.net/. Contact: Beth McGee, CWMTF Water Quality Advisor, Phone: 919-542-5261, E-Mail: beth.mcgee@ncmail.net.

State of Ohio: Water Pollution Control Loan Fund

Description: The Ohio Water Pollution Control Loan Fund (WPCLF), a program of the Ohio Environmental Protection Agency (EPA), provides financial and technical assistance for public wastewater treatment works and nonpoint source water pollution control projects. It awards below market interest rate loans to public and private borrowers. Direct loans are made to most borrowers. Smaller borrowers usually receive indirect loans through a linked deposit program.

The projects that are funded through the WPCLF include the planning, design, and construction of wastewater treatment plants; wastewater treatment plant improvements and expansion; agriculture/silviculture improvements; and stream corridor restoration. The WPCLF staff includes engineers, environmental planners and project coordinators who assist communities in every phase of project development and execution.

Reference for Further Information: WPCLF Website: http://www.epa.state.oh.us/defa/wpclf2.html, Phone: (614) 644-2832. For technical assistance, ask for Pejmaan Fallah, or e-mail Pejmaan at pejmaan.fallah@epa.state.ohio.us.

State of Oregon: Truck Engine Tax Credit

Description: The State of Oregon Department of Environmental Quality (DEQ) offers its Truck Engine Tax Credit to anyone owning a truck in Oregon who purchases a qualifying "cleaner diesel engine" for that truck between 2004 and 2007. The truck must be registered in Oregon. The diesel engine must be purchased in Oregon and have a model year between 2003 and 2007. Additionally, the engine is required to be certified by the U.S. Environmental Protection Agency as emitting oxides of nitrogen at the rate of 2.5 grams per brake horsepower-hour or less. Tax payers may use this credit to reduce their Oregon taxes for any tax year beginning on or after January 1, 2005. The application for the tax credit is a simple one-page form.

Reference for Further Information: For Truck Engine Tax Credit applications and additional information see: http://www.deq.state.or.us/msd/taxcredits/TruckEngine/truckengine.htm. Contact: Oregon DEQ Management Services Division, Tax Credit Program at 503-229-6878. For information about incentives for reducing diesel emissions offered in other U.S. States, and about federal programs, see the Diesel Technology Forum Website at http://www.dieselforum.org/retrofit-tool-kit-homepage/funding-a-program/step-1-identify-possible-funding-sources/.

State of Pennsylvania: Growing Greener Program

Description: In 1999, the Pennsylvania Growing Greener Program was signed into law, providing nearly $650 million to address the state's most pressing environmental challenges. In 2002, the Growing Greener Program's funding was more than doubled. More recently, in 2005, Growing Greener II was signed into law, investing $625 million to extend the Growing Greener Program six years. Funding is provided under the Program for many different types of environmental protection initiatives, including abandoned oil and gas well plugging projects, cleanup and restoration of watersheds, and the construction of new and upgraded water and sewer systems. Counties, local governments, authorities, conservation districts, watershed associations and other nonprofit groups may apply for Growing Greener grants. The Growing Greener Program is the largest single investment to protect the environment in Pennsylvania's history, amounting to $1.2 billion dollars.

Reference for Further Information: To get a list of coordinators, other contacts, applications and other information on Growing Greener and Growing Greener II, visit the Pennsylvania DEP's Website at http://www.depweb.state.pa.us/growinggreener/site/default.asp, or contact Growing Greener at: Phone 717-705-5400, E-mail GrowingGreener@state.pa.us.

State of Rhode Island: Aqua Fund

Description: The Rhode Island Aqua Fund provides grants and loans to fund projects aimed at improving the water quality of Narragansett Bay. The Rhode Island Department of Environmental Management (RIDEM) manages the Aqua Fund. Aqua Fund monies are used to issue grants and loans to cities, towns, universities, nonprofits, governmental agencies, and private agencies. The goal of the Aqua Fund is to remedy existing pollution of Narragansett Bay and to prevent future pollution of the Bay. Funds are issued for projects designed to help prevent pollution to the Bay and its tributaries, such as wastewater treatment projects and urban runoff abatement. Grants are given for up to 90% of the costs of projects under $500,000. Projects with costs exceeding $500,000 may receive grants of up to 50% of total project costs. Projects must be in a location identified as a priority area (see Website). Since the Fund's inception, the Council and RIDEM have awarded over $8.8 million in grants.

Reference for Further Information: Rhode Island DEM, Office of Water Resources Website: http://www.state.ri.us/dem/programs/benviron/water/finance/aqua. Contact: Lisa McGreavy at the Office of Water Resources at lmcgreav@dem.state.ri.us.

When the U.S. Environmental Protection Agency (EPA) was created in the early 1970's its focus was on clean up and control of the most immediate environmental problems. Over the next thirty years, the nation has made huge investments in these pollution control efforts and has realized major reductions in air, water, and land pollution. However, it became increasingly apparent over time that the traditional "end-of-the-pipe" approaches are expensive (increasingly so), not fully effective, and often result in the transfer of pollution between environmental media.

To achieve needed additional improvements to environmental quality, the focus of pollution control efforts must move upstream to prevent pollution before it occurs, and waste must be recycled wherever possible. The Pollution Prevention Act of 1990 recognized that pollution should be prevented or reduced at the source whenever feasible. Consistent with this Act, the U.S. EPA defines pollution prevention as "source reduction." EPA also emphasizes protecting natural resources through conservation and increased efficiency.

The U.S. EPA's environmental management hierarchy for pollution control includes:

    1) source reduction
    2) recycling,
    3) treatment, and
    4) disposal or release.

Preventing pollution offers important economic benefits, as pollution never created does not need to be managed or cleaned up. Recycling means that wastes do not have to be disposed of and raw materials can be conserved. Pollution prevention and recycling have the potential to protect the environment and improve manufacturing efficiency by reducing the use of raw materials.

This section evaluates financing tools which states, communities, and the private sector can use to encourage pollution prevention and recycling. A number of different ways of raising revenues, lowering costs, and influencing behavior are discussed. The tools range from traditional state and federal assistance programs to bold new financial management and investment strategies, programs, and techniques.

1. Assurance and Performance Bonding
2. Demand-Side Management Pricing
3. Deposit-Refund Systems
4. Conservation Pricing for Water Utilities
5. Development Rights Purchases
6. Environmental Self Auditing
7. Full Cost Environmental Accounting
8. Green Investments
9. Liability Assignment
10. Pollution Charges
11. Forest Banks
12. Transit Pass Subsidy Programs
13. Ecotourism
14. Energy Star Program
15. Home Energy Efficiency Mortgages
16. Financial Incentives for Purchasing Hybrid and Alternative Fuel Vehicles
17. Green Suppliers Network
18. U.S. Environmental Protection Agency: Pollution Prevention (P2) Website
19. U.S. Environmental Protection Agency: Natural Gas Star Program
20. U.S. Environmental Protection Agency: Climate Leaders Partnership
21. U.S. Environmental Protection Agency: Green Power Partnership
22. U.S. Environmental Protection Agency: WasteWise Program
23. U.S. Environmental Protection Agency: Pollution Prevention Grant Program

Assurance and Performance Bonding

Description: Mine owners and/or other developers are frequently required under state and federal laws to purchase assurance or performance bonds that cover potential environmental reclamation, restoration, or remediation expenses resulting from their projects. These bonds are repaid to the developers at the time of maturity if the potential damage has not occurred. The bonds are repaid in part if a level of damage below a specified baseline has occurred. The bonds are forfeited if worst-case damages are incurred. In the case that damages occur, these bonds are used to remedy the environmental damages or to compensate injured parties. Assurance and/or performance bonding is required in many states to assure that surface mined areas will be reclaimed and remediated. Also, bonds for site reclamation for coal mines are required under the Surface Coal Mining and Reclamation Act of 1977.

Reference for Further Information: Boyd, James, "Financial Responsibility for Environmental Obligations: Are Bonding and Assurance Rules Fulfilling their Promise?" *Research in Law and Economics,* issue 20, pp. 417-486: 2002.
Boyd, James, "Bonding Requirements for Coal and Hardrock Mines in the U.S." *International and Comparative Mineral Law and Policy: Trends and Prospects,* Bastida, E.; Walde, T.; and Warden, J; editors; New York, Kluwer: 2005. For citations of more of James Boyd's publications, see the Resources for the Future Website at http://www.rff.org/Boyd.cfm#journal.

Demand-Side Management Pricing

Description: Demand-Side Management Pricing, also called Peak Load Pricing, Demand-Responsive Pricing, and Critical Peak Pricing, is a unit pricing structure that is sensitive to the timing of usage (demand) during a utility system's peak hours or peak days. Usage that occurs during these peak periods is charged at a higher rate. Utilities must incur additional capital and operating costs to develop the capacity to meet peak demands. Through demand-side management pricing, these additional costs can be shifted to customers. Such pricing also tends to reduce peak demand by causing system users to reduce their use of the system or at least shift some portion of their usage to non-peak periods. As a result, the utility can "shave" operating costs and stretch existing investment, or reduce future investment in facilities necessary to meet peak period demands. The demand-side management pricing structure is most commonly used by electrical, gas, and communications utilities, and less frequently by water or sewer utilities.

Reference for Further Information: Electric Utility Consultants, Inc. Website: http://www.euci.com/web_conferences/0406-advanced-metering.php. See "Variable Electric Pricing on Tap?: PUC to Consider System that Charges Users More During Peak Midday Hours," by Tribble, Sarah Jane, San Jose Mercury News (CA), p. 1C at: http://www.siliconvalley.com/mld/siliconvalley/14665066.htm.

Deposit-Refund Systems

Description: Deposit-refund systems combine a deposit on a substance or product, paid at the time of purchase, with a refund payable to the consumer when product packaging, or the substance or product itself, is turned in for recycling or proper disposal. Historically, deposit-refund systems have been applied at the state level to glass and aluminum bottles and cans. Deposit-refund systems are now being expanded to include other types of products. For example, in some areas they are being applied to office products, such as photocopy machine toner cartridges and printer inks. The states of California, Maine, and Arkansas are among many states that have established deposit-refund systems to ensure the recycling of lead-acid/automobile batteries.

Reference for Further Information: The Battery Council International outlines state lead-acid battery laws at: http://www.batterycouncil.org/states.html. Publications: Stavins, Robert N.; Harvard University, John F. Kennedy School of Government Research Working Papers Series No. RWP00-004, *Experience with Market-Based Environmental Policy Instruments,* 2001. U.S. Environmental Protection Agency, EPA-240-R-01-001, *The United States Experience with Economic Incentives for Protecting the Environment,* 2001, available at: http://yosemite.epa.gov/ee/epa/eed.nsf/Webpages/SelectReports.html.

Conservation Pricing for Water Utilities

Description: There are a number of different types of full cost pricing used by utilities to encourage water conservation. Six of these pricing structures are described on the U.S. Environmental Protection Agency (EPA) Water and Wastewater Pricing Website. The U.S. EPA defines full cost pricing as "factoring all costs- past and future, operations, maintenance, and capital costs- into prices." The four types of pricing described on the EPA's Website as being most effective in encouraging conservation are increasing block rates, time of day pricing, water surcharges, and seasonal rates.

With increasing block rates, or tiered pricing, charges per unit of water are increased as the amount used increases. The first block is charged at one rate, the next at a higher rate, etc. Time of day pricing is a structure where higher prices are charged during a utility's peak demand periods. Water surcharges are increased rates imposed on water consumption that is considered higher than average. With seasonal rates, prices rise and fall according to water demands and weather conditions, with higher prices generally charged in the summer. Each of these types of pricing qualifies as full cost pricing as long as all costs are recovered through prices.

Many utilities are also assessing surcharges for elevation to cover the increased costs of pumping as power rates increase.

Reference for Further Information: See the U.S. EPA Water and Wastewater Pricing Website at http://www.epa.gov/water/infrastructure/pricing/index.htm.

Development Rights Purchases

Description: The term "Development Rights Purchases" means the purchase of the legal "right" of the owner to develop land for residential or commercial uses. When development rights are purchased for environmental protection purposes, existing land uses are generally maintained. Development rights are often purchased by state and local governments and/or nonprofit groups such as land conservancies. The U.S. Department of Agriculture New York Farm and Ranch Lands Protection Program provides matching funds for the purchase of development rights to keep productive farm and ranch land in agricultural uses.

Development Rights Purchases are like conservation easements, since both entail partial ownership via deed restrictions, contracts or covenants, as opposed to fee-simple transfer of ownership. However, development rights purchases often entail payments to land owners, contrasted to the preferential tax treatment of conservation easements. Buying development rights restricts development whereas conservation easements may require more land management such as soil conservation and plant maintenance to protect water quality and natural habitats. Partial ownership of land through deed restrictions, contracts and covenants is much less costly than outright ownership.

Reference for Further Information: U.S. Department of Agriculture Website: http://www.ny.nrcs.usda.gov/programs/programs/fpp.html.
Environmental Financial Advisory Board, "Protecting America's Land Legacy: Stewardship Policies, Tools, and Incentives to Protect and Restore America's Land Legacy," February 2003, available at http://epa.gov/efinpage/efabpub.htm.

Environmental Self Auditing

Description: The term "environmental self auditing" refers to voluntary efforts by facilities to comply with environmental regulations and improve their environmental performance. Environmental self auditing is carried out through detailed tracking and reporting on a wide range of environmental performance measures. These performance measures include number of Notices of Violations, total emissions, percent of energy usage per unit of output, number of self-identified environmental audit issues compared to the total number of such issues, percentage of issues resolved within established time frames, and percentage of personnel receiving environmental training. The U.S. Environmental Protection Agency's Audit Policy eliminates or reduces civil penalties for violations that facilities disclose as the result of a documented self-audit procedure and correct within 60 days.

Reference for Further Information: U.S. EPA Website: http://www.epa.gov/region5/orc/audits/audit_apil.htm.
Stafford, Sarah, "Does Self-Policing Help the Environment? EPA's Audit Policy and Hazardous Waste Compliance." *Vermont Journal of Environmental Law,* volume 6: 2004-2005, available at: http://www.vjel.org/articles/articles/Stafford11FIN.htm. Also see that article's references section for additional sources and internet links.

Full Cost Environmental Accounting

Description: Full cost environmental accounting is a management accounting method that takes into account all direct and indirect costs, including embedded environmental costs, of a product, process, or activity over its lifetime. The method uses three important concepts: full cost accounting, environmental cost accounting, and life cycle costing. Full cost accounting takes into account historical costs and current costs of the product, process, or activity. Full cost accounting can include potential liabilities due to the handling of hazardous substances or wastes. Environmental cost accounting brings in environmental costs, such as environmental pollution and habitat degradation, and ties them to the product, process, or activity. Finally, life cycle costing identifies the effects of the product, process, or activity at each life cycle stage (raw materials acquisition, manufacturing, use/reuse/maintenance, and recycling/waste management) and assigns those effects monetary values. This accounting method helps incorporate into decisions external costs that are not measured in more traditional accounting methods. Considering these costs is an important step in the pursuit of efficient management and environmental protection.

Reference for Further Information: Association of Chartered Certified Accountants Website: http://www.accaglobal.com/publications/accountingandbusiness/281399. Carnegie Mellon University Pdf files: http://www.ce.cmu.edu/GreenDesign/gd/education/FCA_Module_98.pdf, http://www.ce.cmu.edu/GreenDesign/gd/Research/fca.pdf.
U.S Securities and Exchange Commission Website, SEC disclosure requirements: http://www.sec.gov/rules/final/33-8400.htm.

Green Investments

Description: Green investments are portfolios screened by fund managers to ensure that all capital is invested in companies, financial institutions, monetary funds, and/or other financial entities that have taken clear steps to minimize their environmental impact. The practices of these "environmentally responsible" financial entities may include, for example, use of renewable energy, energy conservation, reduction of solid waste through re-use and recycling, and/or providing products or services designed to help correct an environmental problem. Many managers of green investments require that all companies and institutions they invest in are in compliance with federal, state, and local environmental regulations. These investment portfolios are managed with profit in mind, but are tempered by environmental concerns. Green investments fall under the broader category of socially responsible investments.

Reference for Further Information: Social Investment Forum Website: http://www.socialinvest.org/. A directory of mutual funds categorized as "socially responsible" is available on Co-op America's National Green Pages Online at http://www.greenpages.org/, choose search category "Financial-Mutual Fund Companies" on the drop down menu. For a hard copy of the National Green Pages, call Co-op America at 800-584-7336.

Liability Assignment

Description: Assignment of liability pertains to insurance markets where premiums reflect the relative degree of risk that activities pose to the environment. Premiums send price signals to insurance subscribers creating incentive (i.e., the possibility of lower insurance costs) to prevent pollution, thus reducing their liability. Liability is assigned through common law (negligence) or environmental statutes. The Resource Conservation and Recovery Act (RCRA) program includes a financial responsibility requirement under which treatment, storage, and disposers (TSD's) of hazardous substances must show they can handle the costs of corrective action. This encourages companies to buy insurance to cover the costs of potential damages and provides incentives to avoid releases of hazardous wastes into the environment. Liability standards are also a way to fund remediation activities, i.e., responsible parties are liable for cleanup costs under the Superfund program.

Reference for Further Information: See the U.S. Environmental Protection Agency Website at http://www.epa.gov/compliance/cleanup/superfund/liability.html and http://www.epa.gov/compliance/cleanup/rcra/finance.html.

Pollution Charges

Description: Pollution charges are fees or taxes imposed on polluters based on the amount and/or toxicity of pollution generated. These charges reduce market inefficiencies by discouraging pollution and accounting for the costs of pollution. While the U.S. does not use pollution charges extensively, these charges could be implemented by all levels of government throughout the U.S. Under the Colorado Pollution Prevention Act of 1992, the state levies chemical inventory fees on certain hazardous and extremely hazardous waste generators that exceed the threshold planning quantities under the Superfund Amendments and Reauthorization Act. Russia and Ukraine, and a number of nations in the European Union, including Germany and Great Britain, have considerable experience with pollution charges.

Reference for Further Information: The following reports are available at: http://yosemite.epa.gov/ee/epa/eed.nsf/Webpages/SelectReports.html.
U.S. Environmental Protection Agency (EPA) National Center for Environmental Economics, *The United States Experience with Economic Incentives for Protecting the Environment,* EPA-240-R-01-001, January 2001.
U.S. Environmental Protection Agency (EPA) National Center for Environmental Economics, *International Experiences with Economic Incentives for Protecting the Environment,* EPA-236-R-04-001, November 2004.

Forest Banks

Description: Landowners transfer to the Forest Bank the rights to grow, manage and harvest trees on their land. In exchange, the landowners receive guaranteed annual payments of interest-only on the "principal" (i.e., the assessed market value of the timber at time of deposit in the Bank), a right to withdraw principal revenue on demand, and a guarantee that the timber will remain as forest and will not be developed or clear cut. If the timber on Forest Bank land is harvested, it is done so carefully by the Bank's forester in accordance with a stewardship plan formulated with the landowner. The Forest Bank concept is aimed primarily at non-industrial private landowners.

Reference for Further Information: Dedrick, J.P.; Hall, T.E.; Hull, R.B.; and Johnson, J.E.; "The Forest Bank: An Experiment in Managing Fragmented Forests," *Journal of Forestry,* issue 98, volume 3, pp. 22-25: 2000. Also see the Nature Conservancy Website at http://www.nature.org/ and search on "forest bank" on the Website.

Transit Pass Subsidy Programs

Description: The 1994 Federal Employees Clean Air Incentives Act (Public Law 103-172) provides for the establishment of federal programs to encourage employees to commute by means other than single-occupancy motor vehicles. The purpose of the law is to improve air quality and reduce traffic congestion. Under the Clean Air Incentives Act, the U.S. Environmental Protection Agency (EPA), the U.S. Department of Agriculture (USDA), and other federal government entities implement transit subsidy programs. At the U.S. EPA Headquarters in Washington, D.C., the transit subsidy is provided through a fare card voucher system in partnership with the Washington Area Mass Transit Authority (WAMTA). The fare card vouchers are issued in amounts up to $105 per month for use by participating Agency employees. These vouchers are good for subway fares, bus fares, and any other type of approved public transportation that serves the Washington, D.C. metropolitan area.

Reference for Further Information: Library of Congress Thomas Website: http://thomas.loc.gov/bss/d103/d103laws.html, search on Public Law 103-172 (103rd Congress).

Ecotourism

Description: The International Ecotourism Society defines ecotourism as "responsible travel to natural areas that conserves the environment and improves the well-being of local people." It suggests that people implementing and participating in ecotourism follow these principles: minimizing environmental impact, building environmental and cultural awareness and respect, and providing direct financial benefits for conservation and for local people. In addition, ecotourism activities can advance the three goals of the Convention on Biological Diversity, which are: 1.) conserve biological and cultural diversity by strengthening protected area management systems, 2.) promote the sustainable use of biodiversity, by generating jobs and business opportunities in ecotourism and related business networks, and 3.) share the benefits of ecotourism developments equitably with local communities and indigenous peoples. If carefully targeted and properly implemented, ecotourism offers the real hope of protecting valuable ecosystems while producing a source of revenue for local communities.

Reference for Further Information: International Ecotourism Society Website: http://www.ecotourism.org/. United Nations Environment Programme Website: http://www.uneptie.org/pc/tourism/. Green Globe Sustainable Travel and Tourism Website: http://www.greenglobe.org/.
Conservation International Website: http://www.ecotour.org/xp/ecotour/.

Energy Star Program

Description: Energy Star is a joint program of the U.S. Environmental Protection Agency (EPA) and the U.S. Department of Energy (DOE). The program helps businesses and individuals protect the environment and save money through improved energy efficiency. Energy Star offers tools and resources, such as energy efficiency labeling, to help individuals and households to voluntarily plan and undertake projects that reduce their energy bills and improve home comfort. The program offers a proven energy management strategy for businesses that helps them to measure current energy efficiency, set goals for energy efficiency improvements, and track financial savings resulting from those improvements. In addition, the U.S. EPA recognizes appliances, computers, light bulbs, and entire buildings that meet high standards of energy efficiency with the Energy Star seal. The U.S. EPA has plans to continue to expand the Energy Star program, so that by 2012 it will prevent about 50 million metric tons of carbon equivalent (MMTCE) of greenhouse gases from being emitted each year, equivalent to the emissions of over 30 million vehicles, and will reduce energy bills by about $15 billion annually across the U.S.

References for Further Information: Energy Star Website: http://www.energystar.gov/.

Home Energy Efficiency Mortgages

Description: Home energy efficiency mortgages are a loan category including any mortgage on a new or existing home for which the underwriting guidelines take into account the home's energy efficiency features. Energy efficiency mortgages provide potential and existing homeowners with better than average interest rates and reduced down payments. To qualify for a home energy efficiency mortgage, the homeowner is often required to have his or her home rated by a professional energy rater certified under a national or state accredited home energy rated system (HERS). Both government insured (e.g. FHA, Veteran's Administration) and conventional (e.g. Fannie Mae, Freddie Mac) home energy efficiency mortgages are available. Home energy efficiency mortgages are used to finance technologies such as photovoltaics, solar water and space heating, and energy efficient appliances and systems. The U.S. Department of Energy estimates that an "energy efficient home" can lower home owners' utility bills by 10% to 50%.

Reference for Further Information: Residential Energy Services Network Website: http://www.natresnet.org/resources/lender/default.htm. Energy Star Program Website: http://www.energystar.gov/, information about HERS ratings under "New Homes." Database of State Incentives for Renewable Energy Website: http://www.dsireusa.org/, click on the "Federal Incentives" icon, then go to the "Federal Loan Program" category.

Financial Incentives for Purchasing Hybrid and Alternative Fuel Vehicles

Description: There are many financial incentives available in the U.S. to assist people with purchasing alternative fuel, fuel cell, and hybrid electric vehicles. Some financial incentives offered by states include tax credits, tax deductions, and exemptions from parking fees and the right to use HOV lanes. On the federal level, the Energy Policy Act of 2005 (EPACT) offers consumers and businesses tax credits beginning in January 2006 for the purchase of hybrid electric, alternative fuel, and fuel cell vehicles. The dollar amount of the federal tax credit for the purchase of hybrid electric vehicles varies, and is based on estimates of fuel economy. The federal tax credit for the purchase of fuel cell vehicles is $8,000 until 2009, after which it will be $4,000. The dollar amount of the federal tax credit for the purchase of alternative fuel vehicles is based upon the incremental cost of the vehicle.

Reference for Further Information: See http://www.hybridcars.com/incentives.html and http://www.eere.energy.gov/afdc/laws/incen_laws.html for listings of financial incentives for the purchase of hybrid electric, alternative fuel, and fuel cell vehicles. For information on fuel cell & alternative fuel vehicles, see: http://www.fueleconomy.gov/feg/fuelcell.shtml

Green Suppliers Network

Description: The Green Suppliers Network is a collaborative venture among the U.S. Environmental Protection Agency, the U.S. Department of Commerce's Manufacturing Extension Partnership, and industries participating voluntarily. The Green Suppliers Network works with large manufacturers to engage their small and medium sized suppliers in inexpensive technical reviews focusing on process improvement and waste minimization. These reviews are coordinated through 360vu, a U.S. Department of Commerce program which is the national accounts arm of the Manufacturing Extension Partnership. Through the Green Suppliers Network, participating industries learn manufacturing techniques that help them to increase energy efficiency and optimize resource utilization. Thus, industries participating in the network often increase their profits and reduce their environmental impacts.

Reference for Further Information: Green Suppliers Network Website: http://www.greensuppliers.gov/gsn/home.gsn. Website for 360vu: http://www.360vu.com/prod_serv/index.html.

U.S. Environmental Protection Agency: Pollution Prevention (P2) Website

Description: The U.S. Environmental Protection Agency (EPA) Pollution Prevention (P2) Website has information about state, federal, and private resources and initiatives for pollution prevention, pollution prevention publications, and various approaches to achieving pollution prevention. Under Section 6602(b) of the Pollution Prevention Act of 1990 (Public Law 101-508), Congress established a national policy that pollution should be prevented or reduced at the source whenever feasible. The full text of the Act, and summarized information about it, can be found on the Website.

The federal pollution prevention initiatives described on the Website include programs of the U.S. EPA and other government entities, such as the EPA Pollution Prevention Grant Program, programs funded with government grants, and federal government partnerships with industry. The Website also has a directory, listed under "P2 Resources," of pollution prevention programs in all fifty U.S. States, the District of Columbia, Puerto Rico, and the Virgin Islands. In addition, there are many pollution prevention tips for small businesses, schools, and citizens described on the Website.

Reference for Further Information: U.S. EPA Pollution Prevention Website: http://www.epa.gov/oppt/p2home/index.htm.

U.S. Environmental Protection Agency: Natural Gas Star Program

Description: Natural Gas STAR is a U.S. Environmental Protection Agency (EPA) program consisting of voluntary partnerships that encourage companies across the natural gas and oil industries to adopt cost effective technologies and practices that improve operational efficiency and reduce emissions of methane. This program has many partners who develop a customized implementation plan summarizing how they intend to incorporate the Gas STAR Program into their operations. The U.S. EPA assists Natural Gas Star partners in drawing upon the wealth of partner-provided information, and presents its partners with many opportunities to learn about methane emissions reductions technologies and techniques.

Since the Program began in 1993, Natural Gas STAR partners have eliminated 338 billion cubic feet (Bcf) of methane emissions through the implementation of the Program's core Best Management Practices (BMPs), as well other activities identified by partner companies (referred to as Partner Reported Opportunities). This is the equivalent of removing more than 30 million cars from the road for one year. At the same time, these companies have saved over a $1 billion by keeping more gas in their systems for sale in the market. As of 2004, the companies participating in Natural Gas STAR represent nearly 70% of the natural gas industry in the U.S.

References for Further Information: Natural Gas Star Website: http://www.epa.gov/gasstar/.

U.S. Environmental Protection Agency: Climate Leaders Partnership

Description: Climate Leaders is a voluntary industry-government partnership administered by the U.S. Environmental Protection Agency (EPA). The program works with participating companies, called Climate Leaders Partners, to develop long-term comprehensive strategies for reducing their emissions of greenhouse gases which contribute to global climate change. The Partners use an accounting program that helps them to track their progress in reducing greenhouse gas emissions. Each Partner sets a corporate-wide greenhouse gas reduction goal and inventories their emissions to measure progress. By reporting inventory data to EPA, Partners create lasting records of their accomplishments. Through their participation, Partners identify themselves as corporate environmental leaders and strategically position themselves to stay competitive. The Partners also benefit by receiving technical assistance in the development of their greenhouse gas emissions inventories, and improving their understanding of their greenhouse gas emissions.

Reference for Further Information: U.S. EPA Climate Leaders Program Website: http://www.epa.gov/climateleaders/index.html.

U.S. Environmental Protection Agency: Green Power Partnership

Description: The Green Power Partnership is a voluntary Partnership between the U.S. Environmental Protection Agency (EPA) and organizations interested in buying green power. Through this partnership, EPA supports organizations that buy or plan to buy green power. Green power is a marketing term for electricity that is partially or entirely generated from environmentally preferable renewable energy sources, such as solar, wind, geothermal, biomass, biogas, and low-impact hydro. As a Green Power Partner, an organization pledges to replace a portion of its electricity consumption with green power within a year of joining the partnership. EPA offers credible benchmarks for green power purchases, market information, and opportunities for recognition and promotion of leading purchasers. The goal of the Green Power Partnership is to facilitate the growth of the green power market by decreasing the cost and increasing the value of green power. Due to economies of scale, increased green power generation capacity could lead to substantial cost savings across the board for people purchasing green power. A strong green power market will support new, clean technologies that will reduce the environmental impact of electricity generation.

References for Further Information: U.S. EPA Green Power Partnership Website: http://www.epa.gov/greenpower/aboutus.htm.

U.S. Environmental Protection Agency: WasteWise Program

Description: WasteWise is a U.S. Environmental Protection Agency (EPA) program through which organizations save money by reducing the solid waste and industrial waste that they generate. The program provides free technical assistance to participating organizations, helping them to develop, implement, and measure their waste reduction activities, including recycling and waste prevention. WasteWise is flexible and voluntary, encouraging partners to design waste reduction programs tailored to their individual needs. Through their participation in the program, organizations benefit their bottom line by reducing their purchasing and waste disposal costs. Participating organizations are also assisted in expanding their waste reduction programs through the WasteWise Challenge, an initiative that helps partners to adopt the strategies proven most successful by other partners. In addition, WasteWise provides publicity for participating organizations that are successful in reducing waste, via EPA publications, case studies, and national and regional events.

Reference for Further Information: U.S. EPA WasteWise Website: http://www.epa.gov/wastewise.

U.S. Environmental Protection Agency: Pollution Prevention Grant Program

Description: The Environmental Protection Agency (EPA) Pollution Prevention (P2) Grant Program awards grants to support the establishment and expansion of state pollution prevention programs pursuant to the Pollution Prevention Act of 1990 (Public Law 101-508). The purpose of the P2 Grant Program is to give state programs the capability to assist businesses and industries with identifying better environmental strategies for complying with state and federal environmental regulations. The P2 Grant Program is focused on demonstrating the value of making multimedia pollution prevention an environmental management priority. Eligible recipients include all U.S. states, the District of Columbia, all territories and possessions of the U.S., any agency or instrumentality of a U.S. state, including state universities, and all federally recognized tribes. The majority of P2 grants fund state-based projects. In most cases, grant recipients must contribute fifty percent of the total cost of their project in dollars or in-kind goods and services, but tribal matches can be as low as five percent, if the tribe has a Performance Partnership Grant in place.

Reference for Further Information: U.S. EPA Pollution Prevention Grant Program Website: http://www.epa.gov/oppt/p2home/grants/ppis/ppis.htm. Information on these grants is also available in the *Catalog of Federal Domestic Assistance* at http://12.46.245.173/cfda/cfda.html, search on program number 66.708.

Community-Based Environmental Protection (CBEP) refers to the tailoring of environmental programs and revenue devices to the unique problems and goals of a particular place, such as a watershed, ecosystem, or a community. The CBEP approach is designed to involve localities more intimately in the selection and use of financial mechanisms, maximize the use of scarce resources, enhance the popularity of environmental issues, and more readily involve the private sector in public improvements. CBEP seeks to reflect regional and local conditions, environmental priorities, and economic goals.

CBEP funding approaches are among the most innovative and fastest growing type of financing in this country today. The hallmark of CBEP tools is that most are voluntary and based on the acceptance and active participation of individuals; and they involve both the private and nonprofit sectors in non-traditional roles. Any of the financial tools in this Guidebook can be tailored to local uses and priorities. The unique and highly innovative tools described in this section employ traditional forms of financing as the take-off point. They are designed to reward and encourage environmental protection, unite the public and private sectors in interdependent ways, and bend the forces of the economic marketplace toward these ends. In a sense, CBEP views the public and private sectors as interchangeable, with the government sector supporting the private sector and the latter assuming quasi-government roles.

Parks and recreation, open space conservation, and natural habitat protection are the most popular focal points of community-based environmental protection funding, and they represent the area of greatest volunteerism with the nonprofit sector often taking the lead. Here, communities have adapted some structured funding approaches to meet their needs, such as the formation of special districts, tax increment financing and special tax bonds, and even earmarking a portion of state lottery revenues. Dedicated trust funds, land trusts and conservation easements have been the target of both of public and private financing, including matching funds, special bonds, and corporate and individual donations.

A chief advantage of the majority of community-based environmental financing mechanisms is that they can generate a great deal of enthusiastic and voluntary support without causing much opposition. CBEP financing mechanisms depend upon the enlightened self-interest of individuals, private firms, and governments; and they often result in considerable financial leveraging. They finance long-term protection measures which are not necessarily the target or result of government regulation. The tools in this section capture the spirit, enthusiasm and love that Americans hold for specific regions, valuable natural resources, and the places where they live.

1. Adopt-an-Animal Programs
2. Fundraising Merchandise
3. U.S. Department of Agriculture Farm Service Agency: Conservation Reserve Program
4. U.S. Department of Agriculture Natural Resources Conservation Service: Wetlands Reserve Program
5. Candidate Conservation Agreements with Assurances
6. Capital Improvements Programs
7. Certified Green Buildings
8. Community Foundations
9. Community Supported Agriculture Agreements
10. Conservation Easements
11. Conservation Partnerships
12. U.S. Department of Agriculture Farm Service Agency: Conservation Reserve Enhancement Program
13. Land Trusts
14. Cost-Share Programs for Nonpoint Source Pollution Control
15. Dedicated Government Trust Funds
16. Emissions Trading
17. Lotteries
18. Green Credit Cards
19. Green Energy Partnerships
20. Green Pricing
21. Individual and Corporate Donations
22. Land Reclamation Banks
23. Location Efficient Mortgage
24. Mitigation Banking
25. National Compliance Assistance Centers
26. Nonprofit Organizations
27. Point Source/Nonpoint Source Trading
28. Special Districts
29. Tax Increment Financing

Adopt-an-Animal Programs

Description: Adopt-an-animal programs are used to raise money for zoos, environmental organizations, and environmental programs. Through Adopt-an-animal programs, individuals and corporations pay a fee to adopt particular animals or species. These species are usually officially designated as threatened or endangered under the Endangered Species Act. The individual or corporation is given educational materials on the animal or species they adopt. The Caribbean Conservation Corporation and Sea Turtle Survival League's Adopt-a-Turtle program offers people the option of adopting individual satellite tagged turtles and tracking their movements through their Website. Adopt-an-animal programs can also include publicity and outreach campaigns where experts visit local schools, community centers, and parks to speak about an animal or species and how to help protect it. Additional examples of adopt-an-animal programs include the World Wildlife Fund's program for adopting endangered animals, and the National Zoo's Adopt a Species program.

Reference for Further Information: Caribbean Conservation Corporation and Sea Turtle Survival League Website: http://www.cccturtle.org/member.htm. World Wildlife Fund Website: http://www.worldwildlife.org/, click on "Adoption Center." National Zoo Website: http://nationalzoo.si.edu/Support/AdoptSpecies/.

Fundraising Merchandise

Description: States and localities, as well as the federal government, sell various items and dedicate the revenues to environmental programs. Sometimes revenues are earmarked to site-specific environmental projects. The merchandise sold frequently for this purpose includes auto license plates and stamps. Many states sell special edition license plates to fund environmental conservation activities. The plates are decorated with environmental slogans and designs to show the car owner's support of a particular environmental cause. For example, Maryland and Virginia sell a special "Save the Bay" license plate for $20-$25 each, generating millions of dollars for the Chesapeake Bay Trust and the Chesapeake Bay Restoration Fund. New Hampshire's Moose Plate program supports several grant programs for environmental and cultural conservation. The federal government and most states have duck stamp programs to raise money for waterfowl and wetlands protection projects. These stamps are collector's items.

Reference for Further Information: Chesapeake Bay Trust Website: http://www.chesapeakebaytrust.org/bayplate.html.
Virginia Chesapeake Bay Restoration Fund Website: http://www.state.va.us/dcr/sw/bayfund.htm.
New Hampshire Moose Plate Program Website: http://www.mooseplate.com/overview.html.
The Federal Duck Stamp Program Website: http://www.fws.gov/duckstamps/.

U.S. Department of Agriculture Farm Service Agency:
Conservation Reserve Program

Description: The Conservation Reserve Program (CRP) is a voluntary program for agricultural landowners administered by the Commodity Credit Corporation through the U.S. Department of Agriculture Farm Service Agency. The CRP provides direct assistance payments to eligible applicants to place environmentally sensitive and highly erodible cropland into a 10-15 year contract, i.e., taking it out of crop production. The participant, in return for annual payments, implements a locally approved conservation plan for converting cropland to a less intensive use by planting grasses, legumes, fobs, shrubs, or trees. The CRP seeks to protect the nation's long-term capability to produce food and fiber, reduce soil erosion and sedimentation, create better habitat for fish and wildlife, provide needed income support to farmers, and reduce nonpoint source pollution. The program is leveraged in that it will pay up to 50% of the costs of implementing approved conservation plans.

Reference for Further Information: U.S. Department of Agriculture Farm Service Agency Website: www.fsa.usda.gov/dafp/cepd/crp.htm.

U.S. Department of Agriculture Natural Resources Conservation Service:
Wetlands Reserve Program

Description: The Wetlands Reserve Program (WRP), managed by the U.S. Department of Agriculture Natural Resources Conservation Service, provides technical and financial assistance to eligible landowners to address wetland, wildlife habitat, soil, water, and related natural resource concerns on private lands. The WRP is a voluntary program. The majority of lands enrolled in the WRP are high-risk agricultural lands located in flood prone areas that are restored to wetlands through the program. The goal of the WRP is to achieve the greatest wetland functions and values, along with optimum wildlife habitat, on every acre enrolled in the program. At least 70% of each project area is expected to be restored to natural conditions, and the remaining 30% can be restored to other than natural conditions. There are three options for WRP participants: ten year restoration cost-share agreements, thirty year conservation easements, and permanent easements. There is also a Wetlands Reserve Enhancement Program (WREP) that is administered under the authority of the WRP to enhance delivery of the WRP. The WREP is described in this section of the Guidebook.

Reference for Further Information: U.S. Department of Agriculture, Natural Resources Conservation Service Website: http://www.nrcs.usda.gov/PROGRAMS/wrp/.

Candidate Conservation Agreements with Assurances

Description: Through Candidate Conservation Agreements with Assurances, non-federal property owners commit to implementing conservation measures that protect a specific species considered likely to be listed as threatened or endangered under the Endangered Species Act of 1973 (ESA) in the near future. In exchange for making this commitment, the property owners receive assurances from the Fish and Wildlife Service and the National Marine Fisheries Service that additional conservation measures will not be required and additional land, water, or resource use restrictions will not be imposed on them should the species become listed under the ESA in the future. The assurances are provided in the property owner's Candidate Conservation Agreement, and in an associated enhancement of survival permit issued under section 10(a)(1)(A) of the ESA. Candidate Conservation Agreements with Assurances can create an incentive for landowners to implement conservation measures by providing them with the financial security of protections preventing future restrictions on the use of their land.

Reference for Further Information: Federal Register Notice: http://www.fws.gov/endangered/policy/ccaa.pdf. Environmental Defense Website: http://www.environmentaldefense.org/home.cfm, search on "Candidate Conservation Agreements with Assurances" within the Website.

Capital Improvements Programs

Description: A Capital Improvements Program (CIP) is a planning and financial management process used by public sector agencies for identifying, prioritizing and scheduling planned capital improvements such as construction projects, watershed restoration initiatives, and storm water management projects. CIP's are usually updated and revised on an annual or semi-annual basis. At their most basic, they involve an internal and public review process which results in a prioritized listing and schedule for future capital investments. More sophisticated CIP's also contain a financing element which may consider sources of financing, impacts of facilities on operating costs, and effect on tax rates, debt loads and borrowing limitations. CIPs are used by most medium and large governmental units and public service providers throughout the nation to plan their capital investments, environmental and otherwise. Kansas City; Baltimore County, Maryland; and Florida are examples of jurisdictions with CIP programs.

Reference for Further Information: Kansas City Capital Improvements Management Office Website: www.kcmo.org/cimo.nsf/web/home?opendocument. Baltimore County, Maryland Website: http://www.co.ba.md.us/Agencies/planning/public_facilities_planning/cip.html. Northwest Florida Water Management District Website: http://www.nwfwmd.state.fl.us/rmd/swim/fla_forever_grants/fla_forever_grants.htm.

Certified Green Buildings

Description: Green Buildings are certified by a number of different systems that share the common goals of clean indoor air, resource conservation, and reduced environmental impacts. In the process of Green Building certification, the entire life-cycle of the buildings and their components is usually considered. Certified Green Buildings are typically designed to utilize renewable energy sources and conserve energy. They are often located near public transit and within easy walking distance of stores, schools, and other necessities. In addition, they frequently have water conservation features incorporated into building design and landscaping. Most Certified Green Buildings are also designed to have clean indoor air. That is very important, considering that the air quality in many buildings can be more of a health hazard than the outdoor air in even the most industrialized cities. Certified Green Buildings tend to have low operation and maintenance costs, so they can be a very good investment. Green Building certification can help potential and current home owners to qualify for better mortgage terms.

Reference for Further Information: Smart Communities Network Website: http://www.smartcommunities.ncat.org/buildings/gbintro.shtml. U.S. Green Building Council Website: http://www.usgbc.org/. Build It Green Website: http://www.builditgreen.org/. U.S. Environmental Protection Agency Website: http://www.epa.gov/opptintr/greenbuilding/.

Community Foundations

Description: A community foundation is a federally tax-exempt, non-profit organization that makes grants for charitable purposes in a specific community or region. The funds available to a community foundation are usually derived from many donors and held in an endowment that is independently administered. Income earned by the endowment is used to make discretionary grants meant to build, strengthen and improve the community through environmental protection initiatives and other measures. Although a community foundation may be classified by the Internal Revenue Service as a private foundation under Section 501(c)(3) of the tax code, most are classified as public charities and are thus eligible for maximum tax-deductible contributions from the general public under Section 170 of the code. Community associations' basic appeal to donors is their flexibility. Donors can choose between many different ways of giving tax deductible charitable gifts. The donor can also specify how these donations will be used. This flexibility allows many individuals, through gifts and bequests, to establish permanent endowment funds within one community foundation.

Reference for Further Information: Foundation Center Website: www.fdncenter.org/. Community Foundations of America Website: www.cfamerica.org.

Community Supported Agriculture Agreements

Description: Community Supported Agriculture agreements (CSAs) connect farmers and consumers in mutually beneficial arrangements where consumers buy seasonal "shares" of the harvest at participating farms before the growing season begins. CSAs provide economic benefits to small farms that utilize sustainable agriculture practices, such as the grazing of livestock on large, well vegetated pastures, and reduced pesticide use through Integrated Pest Management. They do this by providing the farmers with a predictable source of income at the beginning of the growing season. They also provide consumers with farm fresh foods that may not be available to them otherwise. CSAs help consumers to have a personal connection with the land on which their food is grown, and they create opportunities for them to communicate with farmers about their environmental practices.

References for Further Information: The Weston A. Price Foundation has many U.S. and international chapters that provide information about CSAs; for chapter contact information see http://www.westonaprice.org/localchapters/locallist.html. The following Websites have CSA directories: Robin Van En Center for CSA Resources: http://www.csacenter.org/, Local Harvest: http://www.localharvest.org/csa/, and Future Harvest-CASA: http://www.futureharvestcasa.org.

Conservation Easements

Description: A conservation easement is a voluntary agreement that allows a private landowner to limit the type or amount of development on his or her property while maintaining partial ownership of the property. The easement is signed by the landowner, who is the easement donor, and by the organization receiving the easement. The receiving organization gains partial ownership of the land. In donating an easement to an organization, the landowner is entrusting that organization with protecting his or her land from future development. Sometimes the organizations receiving the easements pay for them, but in any case acquiring conservation easements is much less expensive than fee-simple purchases. The private landowner gains federal tax benefits if the easement is permanent and is donated "exclusively for conservation purposes." The donor may receive estate and property tax relief as well. Laws in each state vary as to the use and tax implications of private land donations to easements. Easements help to protect agricultural lands and wildlife habitats including forests and water resources.

Reference for Further Information: Byers, Michelle; and Ponte, Karin Marchetti; *The Conservation Easement Handbook,* 2nd Ed: The Land Trust Alliance and the Trust for Public Land, 2005. This and other publications on easements are available through the Land Trust Alliance Website at http://www.lta.org/publications/index.html. The Nature Conservancy Website: www.nature.org. The Trust for Public Land Website: http://www.tpl.org/. Little Traverse Conservancy Website: www.landtrust.org.

Conservation Partnerships

Description: Any arrangement where two or more parties work together to achieve environmental protection objectives, or to help fund environmental initiatives, can qualify as a conservation partnership. There are many different examples of conservation partnerships. Within the U.S. National Park Service, working partnerships are an increasing presence. The National Park Service manages natural heritage areas through collaboration with many diverse organizations. The World Wildlife Fund (WWF) maintains partnerships with organizations that provide funding for WWF and take steps agreed upon with WWF to improve their environmental performance. The Nature Conservancy makes extensive use of conservation partnerships in working with government and the private sector to help protect valuable ecosystems. Conservation partnerships are useful for leveraging funds and pooling resources. In some cases, conservation partnerships may be used for resolving conflicts over land use while protecting ecosystems.

Reference for Further Information: U.S. National Park Service Website: http://www.nps.gov/partnerships/about.htm. Conservation Study Institute Website: http://nps.gov/csi/partnership/practiceDial.htm. World Wildlife Fund Website: http://www.panda.org/abou t_wwf/, search within the Website on "conservation partnerships." The Nature Conservancy Website: http://www.nature.org/partners/.

U.S. Department of Agriculture Farm Service Agency: Conservation Reserve Enhancement Program

Description: The Conservation Reserve Enhancement Program (CREP) is a voluntary land retirement program that helps agricultural producers protect environmentally sensitive land, decrease erosion, restore wildlife habitat, and safeguard ground and surface water. The CREP is an offshoot of the Conservation Reserve Program (CRP), which is the largest environmental improvement program for private lands in the U.S. Both the CREP and the CRP are run by the U.S. Department of Agriculture (USDA) Farm Service Agency. Through combining CRP resources with state, tribal, and private programs, the CREP provides farmers and ranchers with sound financial packages for conserving and enhancing natural resources on farmlands. A specific CREP project begins when a state, tribe, local government, or nongovernmental entity identifies an agriculture-related environmental issue of state or national significance, and works with the Farm Service Agency to develop a project proposal to address that issue.

Reference for Further Information: See the USDA Farm Service Agency Website at http://www.fsa.usda.gov/dafp/cepd/crep.htm. Also see the description of the Conservation Reserve Program in this section of the Guidebook. For additional information about the CREP, contact any county Department of Agriculture Service Center.

Land Trusts

Description: Land trusts acquire and manage ecologically valuable lands on behalf of state and local governments. They are similar to dedicated government trust funds and are funded by a variety of sources, including real estate transfer and property taxes. Land trusts combine management, financing, and planning functions in a single entity. Land management under trusts is often a joint endeavor. Land trusts are highly leveraged, linking public and private funding. Nonprofit land trusts are supported by government monies and private donations. State land trusts are funded by a combination of state (sometimes multi-state), local, and private donations. Revenues from timber harvests, farming, recreation, and other land uses may be rededicated to land trusts. Land purchased for mitigation purposes is often put into trusts. Land acquisition is frequently brokered by nonprofits such as The Trust for Public Land, the Land Trust Alliance, the American Farmland Trust, and The Nature Conservancy. An example of a land trust is the 16,000 acre Sterling Forest acquisition in New York and New Jersey.

Reference for Further Information: Trust for Public Land Website: http://www.tpl.org/. Land Trust Alliance Website: www.lta.org. The Nature Conservancy Website: http://www.nature.org/. American Farmland Trust Website: http://www.farmland.org/. See press release on Sterling Forest at http://www.ny.gov/governor/press/02/may6_02.htm.

Cost-Share Programs for Nonpoint Source Pollution Control

Description: Under cost-share programs, landowners are provided with financial assistance to help them implement Best Management Practices (BMPs) for reducing nonpoint source pollution, which is pollution that is not traced to a specific source. Agricultural Best Management Practices (BMPs), which are conservation measures that prevent water pollution by reducing soil erosion and sedimentation, are frequently implemented with assistance from cost-share programs. Agricultural BMPs include conservation tillage, crop nutrient management, pest management, and conservation buffers. Cost-share programs for nonpoint source pollution control are generally administered and funded at the state level, sometimes leveraging federal, dollars. Delaware, Minnesota, and West Virginia utilize loan money from the federal Clean Water State Revolving Fund to fund their cost-share programs (see the links to the Environmental Protection Agency reports below).

Reference for Further Information: U.S. Environmental Protection Agency Reports: http://www.epa.gov/owmitnet/cwfinance/cwsrf/agfact.pdf and http://www.epa.gov/owm/cwfinance/cwsrf/final.pdf. Virginia Department of Conservation and Recreation Website: http://www.state.va.us/dcr/sw/costshar.htm. Minnesota Cost-Share Program Fact Sheet: http://www.bwsr.state.mn.us/grantscostshare/costshare/factsheet2.html.

Dedicated Government Trust Funds

Description: A dedicated government trust fund is an account set up to receive and disburse revenues for a specific program or activity. The most commonly used methods of raising revenue for dedicated government trust funds include earmarked portions of taxes and fees, referendum bond act dollars, environmental fines and penalties, lotteries, budget surpluses, and private donations. Deposits accrue automatically and usually are available only for the purpose named in the dedication. States and localities throughout the U.S. have dedicated environmental trust funds. Common uses for dedicated environmental trust funds include open space acquisition, habitat restoration, and the operation and maintenance of pollution control facilities. Some examples of state and local dedicated environmental trust funds include: The Nebraska Environmental Trust, South Carolina's Heritage Trust Program, and the Natural Lands Trust Fund in Ocean County, New Jersey. The Superfund Trust Fund and the Nuclear Waste Fund are examples of federal dedicated trust funds.

Reference of Further Information: The Nebraska Environmental Trust Website: www.environmentaltrust.org. U.S. Department of Energy, Energy Information Administration Website: http://www.eia.doe.gov/oiaf/servicerpt/subsidy/excise.html. The Trust for Public Land Website: http://www.tpl.org, search within the site on "land trust fund" to get information about trust funds for land conservation in New Jersey, South Carolina, and other states.

Emissions Trading

Description: Emissions trading programs allow sources of air pollutants to trade pollutants in some fashion, either geographically, over time, or among other sources. Many emissions trading programs incorporate a "bubble" structure. A bubble program treats multiple emission sources as if they were included within an imaginary bubble, allowing existing sources to adjust emissions levels within the bubble as long as an aggregate limit on emissions is not exceeded. "Offset" programs allow new sources to obtain emissions credits from existing sources to offset new emissions. "Banking" programs allow sources to store emission reduction credits for future use or sale. "Netting" programs allow sources undergoing modification to avoid new source review if plant-wide emissions are reduced. In its Clean Air Market Programs, The U.S. Environmental Protection Agency (EPA) employs a flexible emissions trading approach with emissions allowances and caps. This approach is called "allowance trading" or "cap and trade."

Reference for Further Information: U.S. Environmental Protection Agency Website: http://www.epa.gov/airmarkets/trading/basics/index.html. International Emissions Trading Association Website: http://www.ieta.org/ieta/www/pages/index.php. Michigan Department of Environmental Quality Website: http://www.michigan.gov/deq/, click on "Programs" first, then "Emissions Trading."

Lotteries

Description: When operated for the benefit of state or local governments, lotteries generally retain a portion of the revenue from ticket sales, ranging from 10 to 50 percent depending on the game, for dedicated uses such as environmental protection. There are many examples of lotteries being used to raise money for state environmental protection initiatives. The Minnesota Environment & Natural Resources Trust Fund, a permanent source of revenue for environmental and natural resource protection and restoration activities, receives 40% of net Minnesota state lottery proceeds. Colorado has a lottery-funded conservation program called Great Outdoors Colorado. Many other states, including Arizona and Nebraska, use a portion of lottery proceeds to raise money for environmental protection initiatives.

Reference for Further Information: North American Association of State and Provincial Lotteries Website: www.naspl.org. State Environmental Resource Center Website: http://www.serconline.org/conservationfunding/stateactivity.html.
Minnesota Environment & Natural Resources Trust Fund Website: http://www.commissions.leg.state.mn.us/lcmr/trustfund/tfquestion.htm.
Great Outdoors Colorado Website: http://www.goco.org/. Arizona Heritage Fund Website: http://www.azgfd.gov/w_c/heritage_program.shtml. Nebraska Environmental Trust Website: www.environmentaltrust.org.

Green Credit Cards

Description: To issue a "green credit card," a private or nonprofit environmental organization works with a bank or other financial institution. For each green credit card issued, a fixed amount per card and a small percentage of the spending is donated to the organization. Green credit cards are effective fundraising mechanisms for many environmental organizations. Still, it is important to consider that the companies issuing green credit cards may be financing environmentally destructive practices. However, a green credit card is a more "environmentally friendly" alternative to a standard credit card offered by the same company, assuming that the organization supported by the card is using its funds effectively for environmental protection. Many environmental organizations including the National Wildlife Federation, Defenders of Wildlife, and the Nature Conservancy utilize green credit cards for raising revenue.

Reference for Further Information: White, Stephanie, "Buy now, pay later: greening credit cards," *E: The Environmental Magazine,* July-August 2005, available at: http://findarticles.com/p/articles/mi_m1594/is_4_16/ai_n14789145. List of "environment & cause related" credit cards on MBNA America Website: http://www.mbna.com/creditcards/enviro_causes.html.

Green Energy Partnerships

Description: Green Energy Partnerships are groups of businesses and organizations working together to increase their use of "green energy," which includes electricity and heat from renewable energy resources such as wind, solar, and geothermal. Some partnerships also include in their definitions of green power electricity and heat from fuel cells, direct use of landfill gas, and hydroelectric sources they consider to be "low impact." The members of these partnerships provide each other with many forms of assistance including financial savings, networking opportunities, information sources, legislative advocacy, and marketing and promotion. They also may share strategies and lessons learned, and work with power suppliers on marketing initiatives including the design of attractive green electricity products. Based on the theory of economies of scale, as more people purchase green energy with the assistance of these partnerships, it can be expected to become less expensive.

References for Further Information: Clean Energy Partnership Website: http://www.cleanenergypartnership.org. Green Power Market Development Group Website: http://www.thegreenpowergroup.org. U.S. Environmental Protection Agency Green Power Partnership Website: http://www.epa.gov/greenpower/aboutus.htm, also see the "EPA Green Power Partnership" tool in Section 7 of the Guidebook.

Green Pricing

Description: Green Pricing is an option offered by some utilities allowing customers to pay a small premium in exchange for electricity generated from renewable, or "green," energy sources. While there is no universal definition for "green" power, targeted energy sources include solar, wind, geothermal, biomass, ocean power, biodiesel, fuel cells, and some hydroelectric sources considered to be "low impact." The premium charged by utilities covers the additional costs incurred by the electric utility in adding this renewable energy to its power generation mix. Some green pricing programs are accredited through the Center for Resource Solutions' Green Pricing Accreditation Program. Green pricing helps to fund the establishment of new wind farms and other renewable power generation sources. Due to economies of scale, as renewable power sources become more prevalent and more people purchase green power, the cost of these renewable energy sources can be expected to go down.

Reference for Further Information: Center for Resource Solutions Website: http://www.resource-solutions.org/greenpricing.htm. Sterling Planet Website: http://www.sterlingplanet.com/. U.S. Department of Energy Website: http://www.eere.energy.gov/consumer/your_home/electricity/index.cfm/.

Individual and Corporate Donations

Description: Individual and corporate donations, particularly those which are tax deductible, are a popular means for supporting many nonprofit environmental organizations and some government programs. In addition to financial donations, corporate and individual donations may take the form of in-kind payments or special services. Financial donations to governments are often made through line item check-offs on income tax returns allowing taxpayers to earmark a portion of tax refunds for environmental programs. Many nonprofit organizations' operating budgets come largely from individual donations, and to a lesser extent, corporate donations. Private sector corporations and/or foundations frequently match individual financial donations. Donations are frequently used to finance environmental programs that attract significant public interest, such as programs for the conservation of scenic natural habitats, or programs protecting animals that people like to observe on outdoor expeditions, such as whales or sea turtles.

Reference for Further Information: Foundation Center Website: http://fdncenter.org/. Environmental Data Resources, Inc., *Environmental Grant Making Foundations,* 11th ed., 2005, to order see www.environmentalgrants.com, or E-mail orders@environmentalgrants.com.

Land Reclamation Banks

Description: Land reclamation banks are publicly funded or capitalized trust funds that buy contaminated sites and remediate them on behalf of a state or local government. The banks may take title to the land through a number of different means, including purchase, tax foreclosure, and eminent domain. Once the properties are remediated and developed, the bank sells or leases them to generate income for future remediation and development projects. Several cities have used land reclamation banks to clean up and redevelop brownfields. Land reclamation banks combine planning, financing, management, remediation, and redevelopment functions into a single organization. The trust funds for the banks may be financed in several different ways, including tax-increment financing, land transfer taxes, land registration fees, and property sales and leases. Land reclamation banks allow the public or nonprofit sector to assume environmental and financial liability risks that private businesses may be unwilling to assume.

Reference for Further Information: EnviroTools Website: http://www.envirotools.org/financing/privateresources.shtml.

Location Efficient Mortgage

Description: The Location Efficient Mortgage (LEM) combines a three percent down payment, competitive interest rates, and flexible criteria for financial qualification to assist people in buying homes in "Location Efficient Communities." Location Efficient Communities are neighborhoods in which residents can walk from their homes to important amenities such as stores, schools, offices, and public transportation. Location efficiency is a measure of the transportation dollars people can expect to save by living in Location Efficient Communities, based on levels of population and public transit services in their communities. Location Efficient Mortgages are currently available in four metropolitan areas: Seattle, WA; San Francisco, CA; Los Angeles, CA; and Chicago, IL. Location Efficient Communities help to reduce urban sprawl by creating an alternative to communities that depend upon the use of cars. This helps to prevent the need for roads and parking lots that contribute to loss of open space, wildlife habitat degradation, air and water pollution, and stormwater runoff problems.

Reference for Further Information: Location Efficient Mortgage Website: http://www.locationefficiency.com/. Natural Resources Defense Council (NRDC) Website: http://www.nrdc.org/cities/smartGrowth/qlem.asp. Also see: Natural Resources Defense Council, *Solving Sprawl: Models of Smart Growth in Communities Across America:* Island Press, 2003, order through Island Press at http://www.islandpress.org/books/detail.html/SKU/1-55963-432-4.

Mitigation Banking

Description: Mitigation banking is the restoration, creation, enhancement, and, in exceptional circumstances, preservation of wetlands and/or other aquatic resources expressly for the purpose of providing compensatory mitigation in advance of adverse environmental impacts to similar resources. Wetlands Mitigation is required by Clean Water Act (CWA) Section 404 wetlands permits, and by state programs such as Oregon's Removal-Fill Permit Program, to compensate for wetlands removal or fill activities. The purpose of a mitigation bank is to provide for the replacement of the chemical, physical, and biological functions of wetlands and other aquatic resources. The newly established wetland functions are quantified as mitigation credits which are available for use by the mitigation bank sponsor or by other parties to compensate for adverse environmental impacts affecting wetlands.

Reference for Further Information: U.S. Environmental Protection Agency (EPA) Website: http://www.epa.gov/owow/wetlands/guidance/mitbankn.html.
U. S. EPA Wetlands Hotline: 1-800-832-7828. Oregon Department of State Lands Website: http://www.oregon.gov/DSL/PERMITS/wetland_mit.shtml.

National Compliance Assistance Centers

Description: The U.S. Environmental Protection Agency (EPA) operates the National Compliance Assistance Centers in partnership with other government agencies, as well as industries, academic institutions, and environmental groups. The Centers help businesses, local governments, and federal facilities to understand federal environmental laws and regulations and save money via pollution prevention techniques. They achieve this by providing information to those requesting it using means such as Websites, telephone assistance lines, fax-back systems, and e-mail discussion groups. The Centers provide compliance assistance developed primarily for the regulated entities in the following sectors: automotive service and repair, automotive recycling, chemical manufacturing, construction, agriculture, healthcare, metal finishing, paints and coatings, printed wiring board manufacturers, printing, transportation, local governments, federal facilities, and U.S./Mexico border issues. The Centers survey users annually and respondents consistently express a high degree of satisfaction with Center services.

Reference for Further Information: National Compliance Assistance Centers Website: http://www.assistancecenters.net/. U.S. Environmental Protection Agency Website: http://www.epa.gov/compliance/assistance/centers/index.html.

Nonprofit Organizations

Description: Nonprofit organizations, such as foundations and trusts, are defined as non-governmental organizations (NGOs) that accrue no profit to individual members, but spend their resources pursuing specific goals. Nonprofits can be formed for many purposes, including natural lands acquisition, land management, environmental monitoring compliance, research, education, and other activities. They include independent private foundations, operating foundations which make grants to pre-selected organizations, and public foundations funded by government. Many individuals and organizations receive tax deductions for donations to nonprofits; and many governments match private donations to nonprofits. Nonprofits are often able to leverage more monetary donations, volunteer work, resources, and in-kind services than most public agencies. This is due to their tax-exempt status, but also because they provide a focal point that draws attention to the resources they protect and the environmental issues they address.

Reference for Further Information: U.S. Environmental Protection Agency (EPA) Resources for Non-Profit Organizations Website: http://www.epa.gov/epahome/nonprof.htm. Information for nonprofits on the Internal Revenue Service Website: http://www.irs.gov/charities/index.html.

Point Source/Nonpoint Source Trading

Description: In Point Source/Nonpoint Source Trading, a point source of pollution arranges for reduction of nonpoint source pollution discharges in the same watershed in lieu of making more expensive upgrades to its own treatment beyond the minimum technology-based treatment requirements. A number of conditions are necessary for a point source/nonpoint source trading program to achieve ambient water quality objectives. There must be a combination of point sources and nonpoint sources each contributing a significant portion of the total pollutant load in the watershed, and accurate and significant data to establish targets and measure pollution reductions. There must be significant pollutant load reductions for which the marginal cost (cost per pound reduced) for nonpoint source controls are lower than the costs for upgrading point source controls. Under ideal conditions, a trading program will both save money for point source dischargers and improve water quality.

Reference for Further Information: U.S. Environmental Protection Agency Website: http://www.epa.gov/OWOW/watershed/tradetbl.html. Horan, Richard, D, Abler, David G., et. al., "Cost-effective point-nonpoint trading: An application to the Susquehanna River Basin," *Journal of the American Water Resources Association*, April 2002, available at http://www.findarticles.com/p/articles/mi_qa4038/is_200204/ai_n9020966.

Special Districts

Description: A special district is an independent government entity formed to provide and finance governmental services for a specific geographic area, generally at the local level. Residents of special districts pay taxes to finance the improvements from which they will benefit. For example, a sewage special district might tax residents to finance improvements to wastewater treatment services. Special districts issue revenue bonds in a number of states. Examples of special districts include sewer and water districts, storm water management districts, regional solid waste and water resource authorities, regional port authorities, and regional air quality management districts. Local governments use special districts to finance capital facilities independently, relieving the burden on general debt capacity. For example, regional port authorities issue revenue bonds to finance port construction and/or renovation.

Reference for Further Information: Urban Land Institute Website: http://www.uli.org, search the Website on "special districts." Porter, Douglas R.; Lin, Ben C.; Peiser, Richard B; *Special Districts: A Useful Technique for Financing Infrastructure,* Washington, D.C.: Urban Land Institute, 1987.

Tax Increment Financing

Description: Tax increment financing (TIF) provides for the temporary allocation of increased tax proceeds in a designated area generated by increases in assessed property values. In a basic TIF, property assessments are made at a pre-development, or pre-remediation, level in the specified area. Bonds are then issued to finance a portion of the redevelopment or remediation costs. As property values and assessments in the area increase, the municipality uses the added increment in tax revenues to meet the debt service on those bonds. The technique requires the creation of a special district and the maintaining of two separate sets of tax records. Tax increment financing has been used for many years by local governments across the country for a wide variety of economic development projects. It is a particularly effective tool for financing projects that provide measurable specific benefits to select, well defined groups of taxpayers, such as the remediation of a hazardous waste dump near a residential neighborhood.

Reference for Further Information: Illinois Tax Increment Association Website: http://www.illinois-tif.com/. City of San Antonio Economic Development Department Website: http://www.sanantonio.gov/edd/incentive/tax/tif.asp.

The term "brownfield site," as defined by the Small Business Liability Relief and Brownfields Revitalization Act of 2002, with certain legal exclusions and additions, means real property, the expansion, redevelopment, or reuse of which may be complicated by the presence or potential presence of a hazardous substance, pollutant, or contaminant. The definition is available online at http://www.epa.gov/brownfields/glossary.htm. The term "brownfield site" does not include sites listed or proposed for listing on the National Priorities List for cleanup under the U.S. Environmental Protection Agency (EPA)'s Superfund Program. The definition distinguishes brownfields from "greenfields," or undeveloped properties that are not considered to be contaminated.

While many factors can influence economic development decisions, the real or perceived existence of contamination often steers development to greenfields. This promotes the use of land that has never been developed, contributing to urban sprawl, increased traffic congestion, and habitat destruction. It also limits the reuse of brownfields, hurting economic growth in cities. Since many brownfields are located in impoverished communities, such economic decisions may also raise environmental justice concerns. Failure to address the brownfields issue could relegate substantial portions of major cities to environmental and economic decline.

The EPA believes that environmental cleanup is a building block, not a stumbling block, to economic development and that cleaning up brownfields properties must go hand-in-hand with bringing economic vitality to communities. The EPA Brownfields Program and EPA grant programs for the cleanup and redevelopment of brownfields are described in this section. EPA realizes that environmental policy makers must understand the major role of finance in brownfields redevelopment and developers must be educated about environmental requirements.

This Guidebook includes many tools, found mainly in this section but in other sections as well, that can be used to finance brownfields cleanup and redevelopment. Tools in other sections of the Guidebook that can be applied to brownfields cleanup and/or redevelopment include "Land Reclamation Banks" and "Tax Increment Financing" in Guidebook Section 8, and various types of bonds, loans, and grants described in Guidebook Section 2. Additional federal financing mechanisms that can be used for brownfields projects can be found in the Catalog of Federal Domestic Assistance at http://12.46.245.173/cfda/cfda.html.

This section of the Guidebook evaluates financing tools that the federal government, states, communities, and the private sector can use to finance brownfields cleanup and redevelopment. The tools include traditional governmental assistance programs, bold new initiatives that target brownfields sites and disadvantaged communities, innovative private sector arrangements, risk limitation techniques, tax incentives, and use of the Clean Water State Revolving Fund (CWSRF). Most of the financing tools presented here are deeply rooted in local community goals, and include the public and private sectors in a variety of different types of financing arrangements.

1.   U.S. Environmental Protection Agency: Brownfields Program
2.   U.S. Environmental Protection Agency: Brownfields Workforce Development
3.   U.S. Environmental Protection Agency: Brownfields Grants
4.   U.S. Environmental Protection Agency: Clean Water State Revolving Fund Brownfields Loans
5.   State Voluntary Cleanup Programs
6.   U.S. Department of Housing and Urban Development and U.S. Department of Health and Human Services: Round I and II Empowerment Zones
7.   U.S. Department of Housing and Urban Development: Brownfields Economic Development Initiative Grants
8.   National Brownfield Associations
9.   Environmental Risk Management in the Real Estate Industry
10. Industrial Development Funds
11. State Brownfields Programs
12. Land Recycling
13. Real Estate Investment Trusts
14. Tax Abatements
15. Tax Incentives for Brownfields Cleanup and Redevelopment
16. Transferable Development Rights
17. Environmental Insurance
18. Landowner Liability Protections
19. Clean Land Fund
20. Community Development Financial Institutions

U.S. Environmental Protection Agency: Brownfields Program

Description: The U.S. Environmental Protection Agency (EPA) Brownfields Program operates under the authorization of the Small Business Liability Relief and Brownfields Revitalization Act of 2002 (the Brownfields Act). The Brownfields Act, which is an amendment to the Comprehensive Environmental Response, Compensation, and Liability Act (CERCLA), designates the U.S. EPA as the sole agency with authority to award grants to U.S. states, tribes, and territories for brownfields cleanup and redevelopment. Through the Brownfields Program, EPA awards grants to states, tribes, and territories for brownfields assessment and cleanup, capitalization of revolving loan funds, and environmental job training. The U.S. EPA Office of Brownfields Cleanup and Redevelopment administers the Brownfields Program.

Reference for Further Information: U.S. EPA Website, Brownfields Program page: http://www.epa.gov/brownfields/. See "U.S. EPA Brownfields Grants" and "U.S. EPA Brownfields Workforce Development" in this section of the Guidebook. Contact: Sven-Erik Kaiser at the EPA Office of Brownfields Cleanup & Redevelopment at 202-566-2753.

U.S. Environmental Protection Agency: Brownfields Workforce Development

Description: The U.S. Environmental Protection Agency (EPA), other federal agencies, local job training organizations, community colleges, and labor groups have established partnerships to develop long term plans for Brownfields workforce development programs. These workforce development programs, which are often funded with EPA Brownfields Job Training Grants, provide residents of socio-economically disadvantaged communities with job training and employment opportunities with local brownfields cleanup and development initiatives. EPA awards Brownfields Job Training Grants of up to $200,000 to nonprofit organizations, educational institutions, community colleges, tribes, and state and local governments for environmental job training projects that facilitate the assessment, remediation, or preparation of brownfields sites. Through workforce development, EPA facilitates brownfields revitalization while preparing trainees for employment in environmental fields requiring skills in sampling and analysis of hazardous chemicals and remediation of contaminated sites.

Reference for Further Information: U.S. EPA Website: http://www.epa.gov/swerosps/bf/ and http://www.epa.gov/brownfields/job.htm and http://www.epa.gov/brownfields/html-doc/wrkfrc2.htm.

U.S. Environmental Protection Agency:
Brownfields Grants

Description: The U.S. Environmental Protection Agency (EPA) Brownfields Program awards grants to states, tribes, and territories for brownfields assessment and cleanup, capitalization of revolving loan funds, and environmental job training. EPA awards four types of brownfields grants: Assessment Grants, Revolving Loan Fund Grants, Cleanup Grants, and Job Training Grants. Assessment Grants provide funding for grantees to inventory, characterize, assess, and conduct planning and community involvement related to brownfields sites. Revolving Loan Fund Grants provide funding for the capitalization of revolving loan funds that provide sub grants to carry out assessment and/or cleanup activities at brownfields sites. Cleanup Grants provide funding for the cleanup of brownfields sites contaminated by petroleum, hazardous substances, and other pollutants. Job Training Grants are awarded for environmental job training projects that facilitate the assessment, remediation, or preparation of brownfields sites.

Reference for Further Information:
U.S. EPA Website: http://www.epa.gov/brownfields/pilot.htm and http://www.epa.gov/brownfields/facts/rlf_factsheet.pdf. See "U.S. EPA: Brownfields Workforce Development" in this section of the Guidebook regarding Brownfields Job Training Grants.

U.S. Environmental Protection Agency:
Clean Water State Revolving Fund Brownfields Loans

Description: Under Title VI of the 1987 Clean Water Act, states receive federal funding to capitalize Clean Water State Revolving Fund (CWSRF) loan programs. Through CWSRF programs, loans are made to communities to provide low cost financing for a wide range of different projects that protect water quality. Specific brownfields remediation activities that can help prevent contamination of water supplies are eligible uses for CWSRF loans. CWSRF eligible brownfields cleanup activities include removal and remediation of Leaking Underground Storage Tanks (LUSTs) and removal of contaminated soil. Brownfields redevelopment is not an eligible use for CWSRF loans. Any eligible brownfields project must compete with all other water quality projects for a place on the particular state's CWSRF priority funding list.

Reference for Further Information: "U.S. EPA: Clean Water State Revolving Fund" in Guidebook Section 2b. U.S. Environmental Protection Agency Website: http://www.epa.gov/brownfields/ and http://www.epa.gov/owm/cwfinance/cwsrf/.

State Voluntary Cleanup Programs

Description: Voluntary Cleanup Programs are the state and tribal authority, granted by the U.S. Environmental Protection Agency (EPA), to address the environmental and financing problems associated with brownfields and other contaminated properties. Section 128 of the Comprehensive Environmental Response, Compensation, and Liability Act (CERCLA), known as The Small Business Liability Relief and Brownfields Revitalization Act of 2002 (the Brownfields Amendments) codifies EPA's general approach to working with states and tribes on Voluntary Cleanup Programs. EPA enters into non-binding Memoranda of Agreement (MOAs) with states and tribes based on reviews of their Voluntary Cleanup Programs. The purpose of these MOAs is to foster more effective and efficient working relationships between EPA and individual states and tribes regarding their Voluntary Cleanup Programs. In addition, the Brownfields Amendments authorize a noncompetitive $50 million grant program providing funding for states and tribes to establish Voluntary Cleanup Programs.

Reference for Further Information: U.S. EPA Website:
http://www.epa.gov/compliance/cleanup/redevelop/state.html and
http://www.epa.gov/swerosps/bf/pg/fy06_state_guidelines.pdf.

U.S. Department of Housing and Urban Development and U.S. Department of Health and Human Services: Round I and II Empowerment Zones

Description: Empowerment Zones (EZs) are designated geographic areas, usually economically distressed, that are provided with grants for development purposes through the U.S. Department of Housing and Urban Development (HUD) Community Renewal Initiative and the U.S. Department of Health and Human Services. Round I and II EZs may use their grant funding for brownfields cleanup and redevelopment within their borders if those efforts meets several requirements. The requirements are: 1.) the efforts must be consistent with the strategic plan for the EZ, 2.) the efforts must have demonstrated potential to benefit residents of the EZ, and 3.) the governance board for the EZ must approve the expenditures. Round II Empowerment Zones are required to use funds they receive from HUD for brownfields cleanup and redevelopment in conjunction with economic development activities such as job creation or providing sites for businesses. Round I Empowerment Zones, which receive their grant funding through the U.S. Department of Health and Human Services (HHS), are not subject to that requirement. Round III Empowerment Zones are not awarded funding.

Reference for Further Information: U.S. HUD Website:
http://www.hud.gov/offices/cpd/economicdevelopment/programs/rc/.
Contact Phil Graham at HUD: 202-708-6339, ext. 4636. Contact Wally Lumpkin at the HHS: 202-401-5111.

U.S. Department of Housing and Urban Development: Brownfields Economic Development Initiative Grants

Description: The Brownfields Economic Development Initiative (BEDI) is a competitive grant program that the U.S. Department of Housing and Urban Development (HUD) administers to stimulate and promote economic and community development. The BEDI is designed to assist cities with the redevelopment of abandoned, idled, and underused industrial and commercial facilities with real or potential environmental contamination. BEDI grant funds are primarily targeted for use with an emphasis on the redevelopment of brownfields sites in economic development projects. Grant eligible projects are expected to be geared towards creating economic opportunities for low and moderate income persons through the creation or retention of businesses and jobs and increases in the local tax base. BEDI grants are required to be used in conjunction with new Section 108 guaranteed loan commitments. Section 108 is the loan guarantee provision of the HUD Community Development Block Grant (CBDG) Program.

Reference for Further Information: U.S. HUD Website:
http://www.hud.gov/offices/cpd/economicdevelopment/programs/bedi/index.cfm and
http://www.hud.gov/offices/cpd/communitydevelopment/programs/108/.

National Brownfield Associations

Description: National Brownfield Associations (NBA) is an international umbrella organization of national associations that is dedicated to stimulating responsible redevelopment of brownfields. NBA is an educational nonprofit, based in Chicago, with chapters in the United States and Canada. The chapters drive the organization by creating local forums where members can exchange ideas, providing legislative and policy recommendations, hosting events, bridging the communication gap between the public and private sectors, and acting as a trusted source for brownfields information. NBA's mission is carried out through three primary conduits: information, education, and events including conferences, trainings, and meetings in the U.S. and Canada. The guiding principles of NBA include the representation of the diverse interests of brownfields stakeholders nationally and internationally and the encouragement of reuse and redevelopment of brownfields consistent with the environmental and socioeconomic needs of the community. NBA draws its strength from corporate and professional individual members from the public and private sectors in the U.S. and Canada. The magazine Brownfield News is published bi-monthly by NBA, and free copies of it are available via the NBA Website.

Reference for Further Information: National Brownfield Associations Website:
http://www.brownfieldassociation.org/, phone: 773-714-0407.

Environmental Risk Management in the Real Estate Industry

Description: The real estate industry faces financial risks associated with possible contamination of properties that can make participation in brownfields redevelopment unattractive to property owners, purchasers, developers, investors, and lenders. The environmental risks associated with brownfields fall into three categories: cleanup, property value impairment, and personal injury. Parties can reduce or eliminate these risks and make brownfields real estate transactions work using environmental risk management techniques. These techniques either absorb risks, transfer risks between involved parties, or transfer risks to a third party. They include:

1.) *Indemnification* - The seller agrees to cover costs to the purchaser resulting from specific risks.
2.) *Price Adjustment* - The seller reduces the property price to reflect potential contamination risks.
3.) *Self-Insuring* - The purchaser sets aside monies to cover the costs of potential environmental risks.
4.) *Third-Party Insurance* - The seller/purchaser buys insurance to cover potential environmental risks.

Reference for Further Information: Victor O. Shinnerer & Company, Inc. Website: http://www.schinnerer.com/risk_mgmt/real_estate/rmrindex.html#env. Hilb, Rogal & Hobbs Website: http://www.hrh.com/pages/nationalPractices/realEstatePractice.asp. U.S. EPA Website: http://www.epa.gov/swerosps/bf/html-doc/insurnce.htm. See "Environmental Insurance" in this section of the Guidebook.

Industrial Development Funds

Description: Industrial Development Funds (IDFs) are special funds established by state and local governments for the purpose of improving real estate properties to make them suitable for industrial development. These funds are economic development tools that governments use to attract or retain industry. Governments frequently use Industrial Development Funds to fund brownfields cleanup and redevelopment. Industrial Development Funds may be structured as direct pass-through funds or as special purpose revolving funds. These funds are established and replenished through a variety of mechanisms including special property taxes and other taxes, industrial development bonds, surpluses in the controlling government's budget, and proceeds from the sale of real estate and other property. Many states, counties, cities, and towns have laws establishing Industrial Development Funds. These funds may be operated by established government economic development agencies or they may fall under the jurisdiction of special-purpose authorities or corporations. One example of a special-purpose authority is a quasi-governmental non-profit corporation that answers to a government through an appointed board.

Reference for Further Information: National Conference of State Legislatures Website: http://www.ncsl.org/programs/econ/eco-dev.htm, this Website has links to the Websites of the economic development agencies for all U.S. States and Puerto Rico.

State Brownfields Programs

Description: Many states and territories of the United States have brownfields programs that provide financing in the form of loans, grants, and/or bonds for brownfields cleanup and redevelopment. The funding provided to communities through these state and territorial programs is often grant money from the U.S. Environmental Protection Agency (EPA) that is used to award grants and capitalize revolving loan funds. All U.S. states and territories are potentially eligible for EPA funding for brownfields projects. In some cases, states and territories use their own funds for brownfields projects. There are states and territories in every U.S. EPA region with brownfields programs.

Reference for Further Information: See "U.S. EPA Brownfields Program" in this section of the Guidebook. U.S. EPA Website: http://www.epa.gov/brownfields/state_tribal.htm#links & http://www.epa.gov/brownfields/. National Brownfield Associations Website: http://www.brownfieldassociation.org/. Northeast Midwest Institute Website: http://www.nemw.org/reports.htm#brownfields.

Land Recycling

Description: Land recycling revitalizes urban areas and discourages urban sprawl through the cleanup and development of contaminated properties such as brownfields. Identifying brownfields properties and bringing together members of communities, government agencies, financial institutions, and other parties to make brownfields redevelopment work for the benefit of communities is an important part of land recycling. Private land recycling companies and state, county, and municipal land recycling programs help coordinate and finance brownfields assessment and cleanup activities. Many of these companies and programs award loans and grants for brownfields revitalization and other land recycling activities.

Reference for Further Information: Center for Creative Land Recycling Website: http://www.cclr.org/. Pennsylvania Department of Environmental Protection Website: http://www.depweb.state.pa.us/landrecwaste/cwp/view.asp?a=1243&q=462059. American Land Recycling Website: http://www.americanlandrecycling.com/home_1.html. Iowa Department of Natural Resources Website: http://www.iowadnr.com/land/consites/lrp/conLRP.html. City of Phoenix Website: http://phoenix.gov/BROWNFLD/brownfld.html. Wisconsin Department of Natural Resources Website: http://www.dnr.state.wi.us/ORG/caer/cfa/EL/Section/brownfield.html.

Real Estate Investment Trusts

Description: A real estate investment trust (REIT) is an investment corporation that specializes in buying, improving, managing and selling real estate and real estate related assets, including shopping centers, office buildings, hotels, and mortgages secured by real estate. A REIT is essentially a mutual fund that specializes in pooled investments in real estate. The Internal Revenue Code specifies the conditions a company must meet to qualify as an REIT. These conditions include paying 90% of its taxable income to shareholders each year, investing at least 75% of its total assets in real estate, and generating at least 75% of its gross income from investments in or mortgages on real property. REIT dividend earnings can be tax exempt for tax exempt investors such as pension funds. REITs focusing on industrial and commercial real estate could potentially buy, assess, clean up, redevelop, and sell brownfields properties.

Reference for Further Information: National Association of Real Estate Investment Trusts Website: http://www.nareit.com/. U.S. Securities and Exchange Commission Website: http://www.sec.gov/answers/reits.htm. Cornell Law School Website (Internal Revenue Code): http://www.law.cornell.edu/uscode/26/usc_sup_01_26.html.

Tax Abatements

Description: A tax abatement is a temporary moratorium on charging the usual tax rate on a new investment. It may take the form of a full or partial exemption from taxes such as tangible personal property and/or real estate taxes. The exemption lasts for a specific period of time such as five or ten years. The tax abatement granted may be restricted to new development in special designated areas, or it may be targeted on a case-by-case basis to particularly desirable individual development. Tax abatements are individually tailored regarding time and scope to allow the state or local government to calculate the exact cost of the tax change, and thus, the exact tax benefit offered as well. States and communities across the country use various forms of tax abatements to encourage and support economic development. Many additional communities nationwide could direct the use of this type of tax incentive toward brownfields redevelopment and realize substantial environmental economic benefits. If the new development is properly structured and successful, the community tax base will grow at a rate, and to a size, that more than offsets the loss of taxes due to the abatement.

Reference for Further Information: Internal Revenue Service Website: http://www.irs.gov/, search on "abatement." Missouri: North America's Business Center Research Toolbox: http://ded.mo.gov/BDT/topnavpages/Research%20Toolbox/BCS%20Programs/Chapter%20353%20Tax%20Abatement.aspx.

Tax Incentives for Brownfields Cleanup and Redevelopment

Description: There are three basic types of tax incentives offered by federal, state, and local governments: exemptions, credits, and deductions. An exemption provides a release from taxation. Credits provide dollar-for-dollar reductions in taxes owed. Deductions allow certain costs or expenses to be subtracted from income over one year, which is called expensing, or more than one year, which is called depreciation. In addition, tax abatements, which are temporary moratoriums on charging the usual tax rate on new investments, are sometimes used by state and local governments to encourage economic development. A number of states, including New York and Massachusetts, use tax incentives to promote brownfields redevelopment. The federal Brownfields Tax Incentive, a tax deduction encouraging brownfields cleanup and redevelopment, expired on December 31, 2005. The Tax Extender Bill that would reinstate it if passed is pending in Congress.

Reference for Further Information: Fact sheet on New York State tax credits: http://www.empire.state.ny.us/pdf/brownfields/TFtaxsheet011604.pdf.
Massachusetts Department of Environmental Protection Website: http://www.mass.gov/dep/cleanup/certap13.htm.
U.S. EPA Website: http://www.epa.gov/brownfields/html-doc/btaxguid.htm and http://www.epa.gov/brownfields/bftaxinc.htm.

Transferable Development Rights

Description: In transferable development rights programs, property owners are allocated a specified number of transferable development rights (TDRs) in exchange for agreeing to forego or limit development on their land. These property owners are permitted to sell these TDRs to real estate developers, who are permitted to use them to exceed zoning density requirements or other zoning requirements on properties they own in more developed areas. Many states and counties, such as New Jersey and Boulder County, Colorado have transferable development rights programs. In 1998, the Georgia General Assembly passed legislation authorizing local governments to implement transferable development rights programs. Transferable development rights programs might be adapted to encourage brownfields redevelopment, thus protecting other properties from development. Under a brownfields transferable development rights program, developers agreeing to redevelop brownfields could be granted TDRs.

Reference for Further Information:
National Association for Realtors Website: http://www.realtor.org/libweb.nsf/pages/fg804.
Sprawl Guide: http://www.plannersweb.com/sprawl/solutions_sub_tdr.html.
Tools for Quality Growth Website: http://outreach.ecology.uga.edu/tools/tdr.html.
Boulder County, Colorado Website: http://www.co.boulder.co.us/lu/tdr/tdrfaq1.htm.
New Jersey Department of Banking and Insurance Website: http://www.state.nj.us/dobi/pinelandsbank.htm.

Environmental Insurance

Description: Environmental insurance is a tool for managing a party's environmental liability by transferring some of the associated financial risk to another party under the limited provisions of the policy. Essentially, it is an agreement that in return for premium payments, the insured party is provided some protection against unanticipated costs, third party claims, the acts or omissions of other parties, and impairment of property values related to environmental contamination. Environmental insurance is a distinct subset of property and casualty insurance. There can be many different names for environmental insurance policies, but the three most common types that apply to brownfields and other contaminated properties include property transfer insurance, cleanup cost cap or stop loss insurance, and owner controlled insurance. Environmental insurance can be used for real estate/business projects involving the assessment, cleanup and redevelopment of brownfields and other contaminated properties. In brownfields projects, environmental insurance may substitute for indemnities and hold-harmless agreements, lessening the purchaser's need to worry about the seller's financial condition. It may also eliminate the need to report brownfields as environmental liabilities in some cases.

Reference for Further Information: U.S. Department of Housing & Urban Development Website: http://www.huduser.org/publications/econdev/envins.html. U.S. Environmental Protection Agency Website: http://www.epa.gov/swerosps/bf/html-doc/insurnce.htm and http://www.epa.gov/brownfields/insurebf.htm.

Landowner Liability Protections

Description: Under the Comprehensive Environmental Response, Compensation, and Liability Act (CERCLA), known as Superfund, and the Resource Conservation and Recovery Act (RCRA), the owner of a contaminated property can be held liable for coordinating and funding cleanup activities based solely on his or her ownership of the property. However, CERCLA Section 128, known as The Small Business Liability Relief and Brownfields Revitalization Act of 2002 (the Brownfields Amendments) protects landowners meeting statutory criteria from Superfund liability. These liability protections can be applied regardless of whether the landowner's property is a brownfield, a site on the National Priorities List (NPL), or an NPL-caliber site. The Brownfields Amendments also codify the U.S. Environmental Protection Agency (EPA)'s approach to establishing non-binding Memoranda of Agreement (MOAs) with states and tribes based on reviews of their Voluntary Cleanup Programs that address the issue of liability for cleanup of brownfields. See p. 5 of the MOA between EPA and Illinois at http://www.epa.gov/swerosps/rcrabf/mous/illmou.pdf for a discussion of liability.

Reference for Further Information: U.S. EPA Website: http://www.epa.gov/compliance/cleanup/redevelop/landowner.html and http://www.epa.gov/compliance/cleanup/redevelop/state.html.

Clean Land Fund

Description: The Clean Land Fund (the Fund) is a 501(c)(3) nonprofit corporation established in 1999 that is dedicated to remediating and redeveloping targeted brownfields in urban communities in the eastern U.S. Targeted brownfields are sites which 1.) are difficult for the private sector to develop due to marginal economic returns and/or unique environmental conditions, and 2.) are expected to provide public benefits when cleaned up and redeveloped. The Fund provides technical assistance to stakeholders including state and local governments, nonprofit organizations, and private entities by helping them gain access to federal government brownfields funding programs and creating Brownfield Redevelopment Financing Plans. All stakeholders are encouraged by the Fund to participate fully in brownfields redevelopment. The Fund team is made up of experts in the field of brownfields redevelopment. As a nonprofit corporation, the Fund has access to public and philanthropic funding sources to finance brownfields redevelopment projects. The Fund is a member of the National Brownfields Association.

Reference for Further Information: Clean Land Fund Website:
http://www.cleanlandfund.org/, e-mail BillPenn@CleanLandFund.org.

Community Development Financial Institutions

Description: Community Development Financial Institutions (CDFIs) are private sector financial intermediaries with community development, particularly in economically distressed communities, as their primary mission. There are six basic types of CDFIs: community development banks, community development loan funds, community development credit unions, microenterprise funds, community development corporation based lenders and investors, and community development venture funds. The CDFI Fund at the U.S. Department of Treasury works to expand the capacity of CFDIs to provide credit, capital, and financial services to underserved communities throughout the U.S. Examples of CFDIs include the Rural Community Assistance Corporation (RCAC), ShoreBank, The Community Preservation Corporation, and the Community Preservation and Development Corporation. CDFIs could assist with brownfields cleanup and redevelopment in economically distressed communities.

Reference for Further Information: Coalition of Community Development Financial
Institutions (CFDIs) Website: http://www.cdfi.org/. U.S. Department of the Treasury
Community CFDI Fund Website: http://www.cdfifund.gov/.
Rural Community Assistance Corporation Website: http://www.rcac.org/.
ShoreBank Website: http://www.sbk.com/bins/site/templates/splash.asp.
The Community Preservation Corporation Website: http://www.communityp.com/.
Community Preservation and Development Corporation Website: http://www.cpdc.org/.

The environmental goods and services industry (EGSI) is made up of environmental technology companies selling specialized goods and services used for pollution prevention, abatement and remediation. EGSI businesses produce and sell products such as less polluting alternatives to chlorine, bleach, and dry cleaning solvents. In addition, EGSI firms produce and sell recycled paper products, biotechnology products for waste reduction, air pollution reduction equipment, and products for energy conservation. EGSI firms also sell goods to water and wastewater treatment facilities, such as pumps and instruments, and they perform environmental monitoring and testing at those treatment facilities. Many EGSI companies are small businesses. A small business is generally defined in the United States as a business with less than 100 employees.

Small EGSI businesses face many financing challenges that are unique to them, challenges that larger businesses in the industry may not face. In some cases, small EGSI businesses are difficult and costly to capitalize adequately because of weak credit considerations, inadequate experience, poor economies of scale, lack of established markets, and other factors that increase the cost of capital. Small EGSI businesses must compete against much larger engineering and technology companies with market niches, in-house research and development, and years of operating experience in the United States and overseas. Small EGSI businesses can grow and prosper if they access the specialized forms of financing that meet their unique needs. The financial tools presented in this section of the Guidebook include information and analysis to help small EGSI businesses decide which financing mechanisms are right for them.

Section 10 is divided into two subsections of financial tools that can be helpful to small EGSI businesses: Section 10a, "Equity Tools," and Section 10b, "Debt Tools." Section 10a covers equity financing mechanisms such as venture capital, private placements, and related mechanisms such as business plans. Equity financing is money acquired from small business owners themselves or from other investors. Section 10b covers debt mechanisms and related topics such as loans, credit cards, credit analysis, bonds, and leasing. The financial tools presented in Section 10, and a considerable number of tools presented in other sections of this Guidebook as well, could offer great promise to small EGSI businesses, helping them to acquire the financing they need to make significant contributions in the area of environmental protection.

1. Angel Investors
2. Business Plans
3. U.S. Small Business Administration: Active Capital Website
4. Small Business Innovation Research Program
5. Small Business Investment Companies
6. Franchising
7. Investment Forums
8. Investment Networks
9. Joint Ventures
10. Private Placements
11. Public Offerings
12. Strategic Alliances
13. Venture Capital
14. America's Small Business Development Center Network

Angel Investors

Description: An angel investor, or "angel," is an individual who buys into a company, usually in its early stages, as a private investor. Angels provide capital for businesses to start up, usually in exchange for ownership equity. Many angel investors are retired business owners or executives. Angels usually invest their own money. However, a small but growing number of angel investors are organizing themselves into angel networks or angel groups to share research and pool their investment capital. For small businesses and environmental firms, investor angels represent a large source of capital. They are ideal for start-up companies who are too new to qualify for bank loans, expanding companies with growth potential who are too small to attract traditional venture capital, and companies needing only a small amount of money. Professional angel investors seek investments that have the potential to return at least 10 times their original investment within 5 years, through a defined exit strategy, such as a plan for an initial public offering or an acquisition. Most angels invest only in companies that are physically located within 50 miles of where they live or work.

Reference for Further Information: Inc.com, the Daily Resource for Entrepreneurs: http://www.inc.com/magazine/20050701/angels-in-america.html. Gathering of Angels Website: http://www.gatheringofangels.com/. Angel Investor News Website: http://www.angel-investor-news.com/. The vFinance, Inc. Website at http://www.vfinance.com/ has a directory of angel investors.

Business Plans

Description: A business plan is a detailed, formal statement that outlines a company's business goals, the reasons why those goals are believed to be attainable, and the plan for reaching the goals. Business plans encourage managers and employees to think in a more efficient, cost-effective, and strategic manner. The business plan is the most important document that a company will ever produce as a means to obtain financing. Every potential equity investor will want to see a business plan including financial projections and information on what a company intends to do with its monetary and human resources. Business plans are used by most private companies and many non-profit agencies as well. Utilities and other companies providing environmental services can bolster their financial, managerial, and technical capacities to deliver services through the use of business plans. Business plans can also be used by state and federal government offices to spur the optimal in-house provision of environmental services.

Reference for Further Information: DeThomas, Arthur R., Ph.D.; and Grensing-Pophal, Lin; Writing a Convincing Business Plan, Barron's Educational Services, Inc.: 2001, available at http://www.amazon.com/Writing-Convincing-Business-Arthur-DeThomas/dp/0764113992/ref=pd_sim_b_1/002-1123591-2445667.
Purchase software for creating business plans at Jian: http://www.jian.com/.

U.S. Small Business Administration:
Active Capital Website

Description: Active Capital is a Website for entrepreneurs seeking private investment and private investors seeking deals in a secure and protected environment consistent with all investment laws. Active Capital is the replacement for the Angel Capital Electronic Network (ACE-Net), which was created by the U.S. Small Business Administration in 1995. ACE-Net was changed to Active Capital to express more accurately its proactive role in helping small businesses to connect with private capital. Active Capital is the only low-cost internet-based option for registering securities for sale; and it allows registrations of up to $5 million per year. Active Capital's cost effectiveness rests on the fact that it allows entrepreneurial companies seeking equity capital for growth to gain legal access to a nationwide network of investors by answering single sets of questions derived from their business plans. Once accepted into the network, the companies are granted exempt status according to state and federal regulators. This exempt status allows them to go directly to Angels (investors) to acquire capital for their businesses, without having to go through brokers that would require them to pay fees. This exempt status is very beneficial to small businesses with a limited amount of capital.

Reference for Further Information: Active Capital Website: http://activecapital.org/.
U.S. Small Business Administration Website: http://www.sba.gov/. Consult a securities law practitioner for additional information.

Small Business Innovation Research Program

Description: The Small Business Innovation Research (SBIR) Program was established by the Small Business Innovation Development Act of 1982. The purpose of this Act is to strengthen the role of small businesses in federally funded research and development and help develop a stronger national base for technical innovation. All agencies with a federal research and development budget exceeding $100 million are required to participate in the SBIR program. Eleven federal agencies participate. The participating agencies that deal directly with environmental issues include the U.S. Environmental Protection Agency (EPA), the U.S. Department of Agriculture, the U.S. Department of Commerce, the U.S. Department of Energy, the U.S. Department of Transportation, and the National Science Foundation. SBIR businesses are eligible to apply for SBIR financial awards. An SBIR small business is defined as a for profit organization with a maximum of 500 employees that is independently owned and operated and at least 51 percent owned by U.S. citizens or lawfully admitted resident aliens. The U.S. EPA issues annual solicitations for SBIR research proposals in three phases: Phase I, Feasibility Studies; Phase II, Development; and Phase III, Commercialization. The environmental protection related SBIR grant awards included in the U.S. EPA's 2007 Phase 1 solicitation include $3.2 million in small grants for new environmental technologies for purposes such as control and monitoring of air emissions, pollution prevention, and solid waste management.

Reference for Further Information: DARPA's SBIR/STTR Support Center Website: http://www.darpa.mil/sbir/sbir.html. U.S. EPA Website: http://es.epa.gov/ncer/sbir/.

Small Business Investment Companies

Description: Small Business Investment Companies (SBICs) are privately organized and managed venture capital firms licensed by the Small Business Administration's SBIC Program to make equity capital and long-term loans available to small companies. Using private capital and capital borrowed at favorable rates via the federal government, SBICs channel monies to small, fast-growing companies, both new and established. The SBIC program has provided $48 billion in financing to more than 100,000 small U.S. companies since the program's creation in 1958. SBICs specialize in small business financing and have considerable experience/expertise in that area. In addition to loans, they provide expert management assistance to qualifying businesses. Many SBICs make investments or loans in cooperation with other public or private parties. As profit-motivated entities, SBICs expect to share in the success of the small businesses in which they invest. SBICs could be formed for the express purpose of providing capital to small businesses in the environmental goods and services industry, and/or to specific sub-segments of that industry; and they could focus their investments on start-up companies and/or on promoting environmental technology innovation.

Reference for Further Information:
National Association of Small Business Investment Companies Website: http://www.nasbic.org/.
U.S. Small Business Administration Website: http://www.sba.gov/.

Franchising

Description: Franchising is a business partnership where a franchisor licenses a trade name, trademarks, and tried and proven methods of doing business to a franchisee in exchange for a recurring payment, annual fees, and a percentage of gross sales or gross profits. The franchisor provides expertise on a proven business plan to the franchisee. Franchising leverages the franchisor's expertise with the franchisee's money and work. A recent PricewaterhouseCoopers study found that the franchising sector generates 18 million jobs in the United States and yields $1.53 trillion in economic output. There are a growing number of environmental franchises that provide environmental protection related services and/or products, such as leak detection and non-toxic household cleaning supplies. Examples of environmental franchises include Culligan, which provides water purification and delivery services, and Envirospect, Inc., which is one of the many successful franchises performing environmental home inspections. Envirospect, Inc. inspects homes for mold, mildew, allergens, lead, asbestos, and radon, and performs Phase I environmental consulting. The International Franchise Organization (IFA) is a membership organization of franchisors, franchisees, and suppliers whose mission is to protect, enhance, and promote franchising. The IFA's Website provides information on topics including how to franchise a business and how to acquire a franchise.

Reference for Further Information: International Franchise Association Website: http://www.franchise.org/. Gaebler Ventures Website: http://www.gaebler.com/Environmental-Franchises.htm.

Investment Forums

Description: Investment forums are special events, typically serving a state or region, that bring businesses together with investors, economic development officials, and investment intermediaries such as underwriters, venture investment bankers, finance consultants, financial planners, loan brokers, and venture clubs, for their mutual benefit. These forums are sometimes sponsored and attended by small businesses in the environmental goods and services industry. For example, through the Clean Technology Finance and Investment Forums, Cleantech Technology AustralAsia creates a platform for knowledge exchange and commercial interaction for the full range of investors, funds, and companies interested in investing in Cleantech Solutions. Through Cleantech Solutions, Cleantech identifies how it can innovatively finance and deploy Australian clean technology products used to help solve environmental problems and support sustainable development in Australia and internationally. An example of an investment forum is the Eurasian Innovation and Investment Forum that took place in Northern Virginia in March 2007. Earthrise Capital, an investment management company focused on renewable and distributed energy, power conditioning, and power quality, attended that investment forum.

Reference for Further Information: Clean Technology AustralAsia Website: http://www.cleantechnology.com.au/. Eurasian Innovation & Investment Forum Website: http://www.innovationforums.org/virginia. Earthrise Capital Website: http://www.earthrisecapital.com/. Investment Forum Website: http://www.investmentforum.org/.

Investment Networks

Description: Investment networks are business services that match the interests of investors with businesses seeking capital. These networks help new investors to enter into venture capital markets. They also help businesses to attract investment capital. Investment networks provide a confidential way for companies and investors to broaden their range of business contacts. For example, the Capital Network is an investment network that educates and connects entrepreneurs seeking between $50,000 and $4 million who are building fundable businesses toward an exit strategy. Specialized investment networks could be formed for businesses such as recycling companies, pollution prevention enterprises, and environmental remediation technologies firms. The Investors' Circle (IC) Network is an example of an investment network that uses private capital to promote the transition to a sustainable economy. The IC Network is comprised of angel investors, professional venture capitalists, foundations, family offices and others. Recent accomplishments of the IC Network include linking a small venture fund in Vermont with a community development venture fund in Philadelphia and helping an importer of organic cocoa to secure a program-related investment from a major foundation.

Reference for Further Information: See "Venture Capital" in Section 10a of this Guidebook. The Capital Network Website: http://www.thecapitalnetwork.org/. Investors' Circle Website: http://www.investorscircle.net/. Buzzle.com: http://www.buzzle.com/chapters/business-and-finance_investments-and-ventures_networks-and-related-resources.asp.

Joint Ventures

Description: A joint venture is an enterprise formed between two or more parties for the purpose of undertaking economic activity together. The parties each contribute equities to the enterprise; and they then share in the revenues, expenses, and control of the enterprise. A company with limited resources can use a joint venture with a larger corporate partner to exploit its technology in an identifiable market. A joint venture can be for one specific project, such as a research project, or it can be a continuing business relationship. Joint ventures are a common financing technique among firms seeking to bring new products to market. A joint venture can offer a small firm an opportunity to bring technologies or products to market that advance environmental protection goals, such as devices for reducing air pollution.

Reference for Further Information: Consult an attorney on applicable state and federal laws. U.S. Small Business Administration (SBA) Website: http://www.sba.gov/, contact SBA at answerdesk@sba.gov, or at 1-800-U-ASK-SBA (1-800-827-5722) with questions about joint ventures or for referral to the nearest Small Business Development Center. JustinMichie.com: http://www.justinmichie.com/free_articles/joint_venture_marketing_on_the_internet.php.

Private Placements

Description: A private placement is the direct sale of a limited number of shares of stock in a company to a relatively small number of pre-selected buyers, often institutional investors. The most common type of private placement is the limited partnership. A typical limited partnership involves one general partner who holds full authority for all business decisions, and a number of limited partners who serve as angel investors. Private placements are used by new companies that are not likely to attract a single investor to come up with the entire amount of money that they need. They are also used by established businesses that want to acquire money without the scrutiny, expense, and paperwork involved with public offerings. Businesses providing environmental protection related services and products could raise funds using private placements. Private placements can normally be used to raise between $100,000 to $1,000,000, and sometimes more. Successful private placements require very good business plans.

Reference for Further Information: See "Angel Investors" in this section of the Guidebook. Blechman, Bruce Jan; and Levinson, Jay Conrad, Guerrilla Financing: Alternative Techniques to Finance Any Business, Houghton Mifflin Company, Boston, Massachusetts: November 1991, available at http://www.amazon.com/Guerrilla-Financing-Marketing-Bruce-Blechman/dp/0395522641.
About Website: http://sbinformation.about.com/cs/creditloans/a/prplacemt.htm. "Private Placement Letter" Website: http://www.privateplacementletter.com/protected/current_issue.cfm.

Public Offerings

Description: A public offering (PO) is the selling of a company's shares to the public in the form of stock. Public offerings require registration with the Securities and Exchange Commission (SEC) and with state regulators. The first time a company sells shares to the public is known as an initial public offering (IPO). Small Corporate Offering Registrations allow smaller companies raising less than $5 million to be exempt from some regulations. A growing number of companies that focus on environmentally-related technologies are selling their stock in public offerings. For example, CECO Environmental Corporation, a producer of air pollution control and industrial ventilation systems, announced on April 12, 2007 that about 3.3 million shares of its common stock will be sold in a public offering. Public offerings allow companies with well-established records of successful performance to raise money for investment in product development, marketing and business expansion. Through public offerings, companies normally raise at least $5 million; and they sometimes raise hundreds of millions of dollars.

Reference for Further Information: U.S. Securities and Exchange Commission Website: http://www.sec.gov/. IPOHome Website: http://www.ipohome.com/default.asp. "CECO Announces Public Offering," *Associated Press Business News,* 2007, available at http://news.moneycentral.msn.com/category/topicarticle.aspx?feed=AP&Date=20070412&ID=6737829.

Strategic Alliances

Description: A strategic alliance is a partnership formed between two or more parties to pursue a set of agreed upon goals. Strategic alliances are often formed to meet critical business needs. There may be some exchange of equity in these corporate partnerships, but both companies continue to operate as independent entities. Strategic alliances are often used in research contracts and marketing and licensing agreements. These alliances can also be used for financing, product distribution, and many other business activities. A strategic partner may provide financing structured as an investment, a loan, prepayment for work to be performed, or as an exchange or sharing of resources such as space, personnel, and equipment. Many small businesses form strategic alliances with larger companies to get their products or services to the market faster. An example of a potential strategic alliance with environmental protection benefits is car manufacturers forming an intra-industry alliance to develop electric car technology. Businesses can rely on the Association of Strategic Alliance Professionals, which is a global professional association, to help make strategic alliances happen by providing a forum for exchange of information on topics such as best practices, resources, and opportunities.

Reference for Further Information: Association of Strategic Alliance Professionals Website: http://www.strategic-alliances.org/.

Venture Capital

Description: Venture capital is a type of private equity capital typically provided by professional, institutionally-backed outside investors to new businesses with high growth potential. Individuals who make these types of investments are called venture capitalists. Venture capital funds are pooled investment vehicles (often partnerships) that invest in enterprises considered too risky for standard capital markets or bank loans. Professionally managed venture capital firms are generally private partnerships or closely-held corporations. Some examples of venture capital firms with an environmental focus include Alyra Renewable Energy Finance, LLC and EnerTech Capital. Alyra provides financial advisory services to businesses in the renewable energy sector. EnerTech helps entrepreneurs create companies that deliver profitable, differentiated energy technologies such as wind power and solar power.

Reference for Further Information: See the EcoBusinessLinks Environmental Directory for a directory of green venture capital firms:
http://www.ecobusinesslinks.com/green_venture_capital.htm.
See the vFinance, Inc. Website for a directory of venture capital firms:
http://www.vfinance.com/.
National Venture Capital Association Website: http://www.nvca.org/def.html.
EnerTech Capital Website: http://www.enertechcapital.com/.
Alyra Renewable Energy Finance, LLC Website: http://www.alyra.net/. .

America's Small Business Development Center Network

Description: America's Small Business Development Center (SBDC) Network, also called the SBDC program, is a comprehensive small business assistance network serving the United States and its territories. The SBDC Network is a partnership that includes Congress, the U.S. Small Business Administration, the private sector, and the colleges, universities, and state governments that manage SBDCs. The mission of the SBDC Network is to help new entrepreneurs realize their dreams of business ownership, and assist existing businesses to remain competitive in the complex marketplace of an ever-changing global economy. The services provided by the SBDC Network include free face-to-face business consulting and at-cost training on writing business plans, accessing capital, marketing, regulatory compliance, international trade, and more. The SBDC Network provides its services to over 1.3 million small business owners and aspiring entrepreneurs each year. SBDC in-depth clients generate $2.82 in new federal tax revenue for every $1 spent by the U.S. government on the SBDC program.

Reference for Further Information: America's Small Business Development Center Network Website: http://www.asbdc-us.org/. U.S. Small Business Administration (SBA) Website:
http://www.sba.gov/services/financialassistance/sbaloantopics/prequalification/index.html, SBA phone # 1-800-U-ASK-SBA (1-800-827-5722).

1. U.S. Department of Agriculture Rural Development: Intermediary Relending Program
2. Bank Financing
3. U.S. Small Business Administration: Prequalification Loan Program
4. U.S. Small Business Administration: Certified Development Company/504 Loan Program
5. U.S. Small Business Administration: Basic 7(a) Loan Program
6. U.S. Small Business Administration: Microloan Program
7. Small Business Administration: CAPLines Loan Program
8. Community Reinvestment Act
9. Convertible Debt
10. Credit Analysis
11. Credit Cards
12. Export-Import Bank of the United States: Environmental Exports Program
13. Foundations: Program-Related Investments
14. Leasing
15. Mezzanine Financing
16. Microcredit
17. National Consumer Cooperative Bank
18. National Credit Union Administration: Community Development Revolving Loan Fund
19. Accounts Receivable Financing
20. Surety Bonds
21. U.S. Department of Treasury: Community Development Financial Institutions Fund

U.S. Department of Agriculture Rural Development: Intermediary Relending Program

Description: The U.S. Department of Agriculture (USDA) Rural Development Intermediary Relending Program (IRP) was established for the purpose of alleviating poverty and increasing economic activity and employment in rural communities. The IRP provides loans to local organizations that act as intermediaries for the establishment of revolving loan funds. These revolving loan funds are used to help finance businesses and economic development activities for the purpose of creating or retaining jobs in disadvantaged and remote communities with populations of 25,000 or less. The intermediaries are encouraged to work in accord with state and regional strategies, and in partnership with other public and private organizations that can provide complimentary resources. Intermediary recipients of these loans include private non-profit corporations, public agencies, Indian tribes, and cooperatives. An intermediary may borrow up to $2 million for its first financing and up to $1 million for any financing thereafter. Total aggregate debt for any intermediary is capped at $15 million. Environmental protection related initiatives that are financed with these loans include purchases of easements to protect land from development and pollution control and abatement projects.

Reference for Further Information: U.S. Department of Agriculture Website: http://www.rurdev.usda.gov/rbs/busp/irp.htm. Organizations must contact their USDA Rural Development State Offices to apply for loans through this program. To locate the Website of your USDA Rural Development State Office, see http://www.rurdev.usda.gov/recd_map.html.

Bank Financing

Description: Banks frequently extend credit to businesses and individuals through loans and lines of credit. There are two types of bank loans: secured and unsecured. For secured loans, the borrower puts up collateral such as real estate, stocks, business assets, or something else of value that the bank can take if the borrower defaults on the loan. Unsecured loans are based on the credit of the borrower, do not require collateral, and often have higher interest rates than secured loans. Lines of credit are open accounts that can be drawn upon as needed up to their limit. These accounts have higher interest rates and are more expensive compared to most loans. Small business lines of credit generally have lower limits than loans. Bank financing is a good option for established businesses that are profitable and have good credit records. Many businesses providing environmental protection related services and products rely on bank financing. The American Bankers Association (ABA), the largest banking trade association in the United States, provides training and other resources to assist businesses with financing. The ABA Website provides information on topics such as survey reports comparing banks to their peers, consumer education, and the performance of farm banks in providing agricultural loans.

Reference for Further Information: American Bankers Association Website: http://www.aba.com/default.htm.

U.S. Small Business Administration:
Prequalification Loan Program

Description: The U.S. Small Business Administration (SBA) Prequalification Loan Program uses intermediary organizations to assist prospective borrowers in developing viable loan application packages and securing loans. Through this program, loans are awarded to low income borrowers, disabled business owners, new and emerging businesses, veterans, exporters, and rural and specialized industries. The role of the intermediary organization is to work with the loan applicant to make sure the business plan is complete and that the application is eligible and has credit merit. The intermediaries send the loan applications to the SBA when they are satisfied that they have a chance for approval. If the SBA approves the loan application, the intermediary then helps the borrower locate a lender offering the most competitive rates. Unlike the Small Business Development Centers serving as intermediaries for this loan program, for profit organization intermediaries do charge a fee. The maximum loan amount offered through this program is $250,000. Businesses providing environmental protection related services and products could potentially be awarded these loans.

Reference for Further Information: U.S. Small Business Administration (SBA) Website: http://www.sba.gov/services/financialassistance/sbaloantopics/prequalification/index.html. Call the SBA at 1-800-U-ASK-SBA (1-800-827-5722) to locate a prequalification intermediary.

U.S. Small Business Administration:
Certified Development Company/504 Loan Program

Description: The Small Business Administration (SBA)'s Certified Development Company/504 Loan Program is a financing tool for economic development within a community. Under the 504 Program, nonprofit corporations called Certified Development Companies (CDCs) work with the SBA and private-sector lenders to provide long-term, fixed-rate financing to small businesses for substantial fixed assets, such as land and buildings. Section 504 of the Small Business Investment Act requires the SBA to sell debt instruments called debentures to investors. A project financed by the 504 Program generally includes a loan secured with a senior lien from a private-sector lender covering up to 50% of the project costs, a loan secured with a junior lien from a CDC (backed by a 100% SBA-guaranteed debenture) covering up to 40% of the cost, and a contribution of at least 10% equity from the small business being helped. The maximum SBA debenture is $1.5 million when meeting the job creation criteria or a community development goal and $2 million when meeting a public policy goal. Generally, a business is required to create or retain one job for every $50,000 provided by the SBA under this program. An example of an environmental protection initiative that the 504 Program could potentially finance is the modernization of buildings through making them more energy efficient.

Reference for Further Information: U.S. Small Business Administration (SBA) Website: http://www.sba.gov/services/financialassistance/sbaloantopics/cdc504/index.html and http://www.sba.gov/tools/resourcelibrary/lawsandregulations/tool_lawsreg_Smallbusinvst.html. SBA phone: 1-800-U-ASK-SBA (1-800-827-5722).

U.S. Small Business Administration:
Basic 7(a) Loan Program

Description: The U.S. Small Business Administration (SBA) Basic 7(a) Loan Program is the most basic and most used type of SBA's business loan programs. The program is regulated under Section 7(a) of the Small Business Act (Public Law 85-536, as amended). To be considered for financing under the Basic 7(a) Loan Program, businesses must show that they lack the internal resources (business or personal) to provide financing and demonstrate that they cannot get financing on reasonable terms via any other lending channels. Basic 7(a) loans are only made available on a guaranty basis and they have a maximum loan amount of $2 million. The SBA's maximum exposure for 7(a) loans is $1.5 million. Thus, if a business is awarded a 7(a) loan of $2 million, the maximum guaranty to the lender is $1.5 million. Most American banks participate in the 7(a) program, and there are some non-bank lenders as well. Interest rates on 7(a) loans may be fixed or variable, and they vary depending upon the size and terms of the loan. Basic 7(a) loans could be used for environmental protection purposes such as the purchase of equipment needed to meet Clean Air Act or Clean Water Act requirements.

Reference for Further Information: U.S. Small Business Administration (SBA) Website: http://www.sba.gov/services/financialassistance/sbaloantopics/7a/index.html & http://www.sba.gov/regulations/sbaact/sbaact.html.
To locate a lender, call the SBA at 1-800-U-ASK-SBA (1-800-827-5722).

U.S. Small Business Administration:
Microloan Program

Description: The U.S. Small Business (SBA) Administration Microloan Program provides very small loans to start-up, newly established, or growing small businesses. The SBA Microloan Program is regulated under Section 7(m) of the Small Business Act (Public Law 85-536, as amended). Under this program, the SBA awards funds to nonprofit community based lenders (intermediaries) which, in turn, make loans to eligible borrowers in amounts up to a maximum of $35,000. The average sized SBA microloan is about $13,000. The maximum term permitted for microloans is six years. Interest rates on these loans vary, depending upon the intermediary lender and the costs to the intermediary from the U.S. Treasury. The interest rates are generally between 8% and 13%. Many different types of small businesses can meet SBA eligibility requirements and qualify for a microloan. Microloans can provide needed capital to start or expand businesses in the environmental goods and services industry.

Reference for Further Information: U.S. Small Business Administration (SBA) Website: http://www.sba.gov/services/financialassistance/sbaloantopics/microloans/ and http://www.sba.gov/regulations/sbaact/sbaact.html. Call the SBA at 1-800-U-ASK-SBA (1-800-827-5722) or see the listing on the SBA Website to locate an SBA microlending intermediary.

U.S. Small Business Administration: CAPLines Loan Program

Description: CAPLines is the umbrella program under which the U.S. Small Business Administration (SBA) helps small businesses meet their short-term and cyclical working capital needs. The CAPLines loan program is regulated under Section 7(a) of the Small Business Act (Public Law 85-536, as amended). The following five short-term working capital loan programs for small businesses are offered under the CAPLines umbrella: 1.) Seasonal Line: advances against anticipated inventory and accounts receivable help, 2.) Contract Line: finance direct labor and material costs of performing assignable contracts, 3.) Builders Line: finance direct labor and material costs for construction and renovation projects, 4.) Standard Asset-Based Line: revolving lines of credit for firms unable to meet credit standards associated with long-term credit, and 5.) Small Asset-Based Line: revolving lines of credit of up to $200,000. The SBA provides a maximum loan amount of $2 million for all CAPLines loans except for the Small Asset-Based Line. CAPLines loans can help small businesses in the environmental goods and services industry to meet payroll, acquire building materials, and pay other expenses.

Reference for Further Information: Small Business Administration (SBA) Website: http://www.sba.gov/smallbusinessplanner/start/financestartup/SERV_CAPLINES.html and http://www.sba.gov/regulations/sbaact/sbaact.html.
SBA phone #: 1-800-U-ASK-SBA (1-800-827-5722).

Community Reinvestment Act

Description: The Community Reinvestment Act (CRA) of 1977 encourages depository institutions to help meet the credit needs of the communities in which they operate, including low and moderate income neighborhoods, consistent with safe and sound banking operations. The CRA requires that each insured depository institution's record in helping meet the credit needs of the community in which it is located be evaluated periodically. The Office of the Comptroller of Currency revised its regulations on implementation of the CRA in 1995. Under the 1995 regulations, lenders subject to the CRA can claim community development loan credits for loans made to help finance the environmental cleanup or redevelopment of industrial sites when it is part of an effort to revitalize the low and moderate income communities in which these sites are located. The 1995 regulations are designed to encourage economic activity in urban areas and they make the financing of industrial property development more attractive to large lenders by providing CRA credit while aiding the communities in which they operate.

Reference for Further Information: Federal Financial Institutions Examination Council (FFIEC) Community Reinvestment Act Website: http://www.ffiec.gov/cra/default.htm. U.S. Department of the Treasury Comptroller of the Currency Website: http://www.occ.treas.gov/. U.S. Environmental Protection Agency Website: http://www.epa.gov/swerosps/bf/html-doc/cra.htm. See Section 9 of this Guidebook for more information on Brownfields.

Convertible Debt

Description: Convertible debt is a loan instrument obtained from a potential capital investor that can be converted to equity in the form of stock ownership under certain circumstances at a future date. The lender earns interest on the loan until it is either paid off or the conversion option exercised. The decision regarding whether or not to convert the debt to equity usually rests with the lender. Convertible debt offers lenders and investors flexibility by permitting them to control some of the risks and rewards of their investments. Lenders have the option of negotiating down the future equity offer in return for higher interest returns on the debt up-front. Convertible debt is usually more expensive than straight business loans. The most common forms of convertible debt issued by businesses are securities such as debentures, bonds, and stock options. Many businesses, including small companies, use convertible debt as a means of raising capital to finance business start-ups and/or expansions. Businesses in the environmental goods and services industry could evaluate the option of using convertible debt to raise capital.

References for Further Information: Entrepeneur.com Website: http://www.entrepreneur.com/money/financing/startupfinancingcolumnistasheeshadvani/article159520.html. Business Owners' Idea Café Website: http://www.businessownersideacafe.com/financing/convertible_debt.php.

Credit Analysis

Description: Credit analysis is the process of measuring a credit applicant's ability and willingness to repay debt. It is an assessment of the probability that the applicant will pay on time and in full. The standard framework for credit analysis is "The Four C's" of credit: 1.) creditworthiness: a measure of a borrower's willingness to make timely payments on debt, 2.) collateral: the value of any assets pledged as security for a debt, 3.) capacity: an assessment of a borrower's ability to repay a debt, and 4.) capital: a measure of how much cash (or assets readily converted into cash) that the borrower has to make a down payment, cover closing costs, and handle other incidental expenses. Credit analysis is done through a review of the applicant's credit history, financial statements, and records on file with credit reporting companies. Credit analyses are performed routinely by banks, savings and loans, other financial institutions, investors, businesses, and individuals. Knowledge of the credit analysis process can help credit applicants to manage their finances in such a way as to improve their credit ratings.

Reference for Further Information: Contact the credit department of any bank, savings and loan, finance company, or other lending institution. Nationwide Advantage Mortgage Website: http://www.nationwideadvantage.com/siteinfo/nw/content/jsp/YourCredit-TheFourCs.jsp. Golan, Jonathan, The Bank Credit Analysis Handbook: A Guide for Analysts, Bankers, and Investors, John Wiley & Sons, New York: 2001, available at http://www.amazon.com/Bank-Credit-Analysis-Handbook-Investors/dp/0471842176.

Credit Cards

Description: Individuals, corporations, and other parties frequently use credit cards for purposes such as covering travel expenses and purchasing office supplies. Many entrepreneurs and small businesses use credit cards to help finance business start-ups. When used responsibly, credit cards represent a good source of modest amounts of debt capital. Credit cards are particularly good for meeting any expense that can be repaid within the interest free grace period, which is generally 30 days, or during the interest free introductory periods of several months that some cards have. Credit cards are often used for isolating and/or itemizing different business expenses through the designation of specific cards for specific types of expenses. Businesses and individuals could use credit cards for short-term financing of many different types of environmental protection initiatives. Of course, care must be taken on the part of credit card users to use good financial planning so that they enhance, rather than impair, their credit ratings. The Credit Infocenter and Smartaboutmoney.org provide useful information on topics such as managing your money and reading a credit report to help people use credit cards responsibly.

Reference for Further Information: See "Credit Analysis" in this section of the Guidebook. CreditInfocenter.com: http://www.creditinfocenter.com/. SmartAboutMoney.org: http://www.smartaboutmoney.org/nefe/pages/content.asp?page=1720.

Export-Import Bank of the United States: Environmental Exports Program

Description: The Export-Import Bank of the United States (Ex-Im Bank) is the official export credit agency for the United States. The mission of the Ex-Im Bank is to assist in financing the export of U.S. goods and services to international markets. The Ex-Im Bank's Environmental Exports Program (EEP) provides enhanced levels of financial support for exports of a broad range of environmentally beneficial products including renewable energy products. Enhanced support under EEP is made available to finance exports needed for environmentally beneficial projects such as pollution cleanup initiatives and renewable energy projects, and for exports of products and services specifically used to prevent, abate, control, or mitigate air, water and ground contamination or pollution, or to provide protection in the handling of toxic substances and wastes. Financing mechanisms offered through the EEP include: 1.) short-term environmental export insurance policies, 2.) a working capital guarantee enabling small and medium sized businesses to access funds to purchase materials and goods for export and to meet other export related financing needs, and 3.) enhanced medium-term insurance and guarantees and long-term loans and guarantees for export transactions.

Reference for Further Information: Export-Import Bank of the United States Website: http://www.exim.gov/ & http://www.exim.gov/products/special/environment.cfm.

Foundations: Program Related Investments

Description: A program related investment is a loan or other investment made by a foundation to a business or nonprofit organization for a project related to the foundation's stated purpose and interests. Program related investments often take the form of loans with limited or below-market interest, loan guarantees, equity investments, asset purchases, linked deposits, purchases of stock, or other forms of financial support. Foundations make program related investments mainly to maximize the impact of their programs, provide an alternative form of financing when grants are inappropriate or insufficient, and, in the case of loans, recycle dollars to increase funding availability. Program related investment dollars are often used to provide capital to intermediary organizations, such as revolving loan funds and development banks, which in turn lend funds to development agencies and service providers. Foundations that make program-related investments include the Ford Foundation, the John D. and Catherine T. MacArthur Foundation, and the Met Life Foundation. Program related investments can be a source of low cost debt or equity capital to support an organization's environmental projects.

Reference for Further Information: Foundation Center Website: http://foundationcenter.org/. John D. and Catherine T. MacArthur Foundation Website: http://www.macfound.org. Ford Foundation Website: http://www.fordfound.org/program/community.cfm. MetLife Foundation Website: http://www.metlife.com/Applications/Corporate/WPS/CDA/PageGenerator/0,4132,P263,00.html.

Leasing

Description: A lease is a rental contract granting the right to use or occupy personal property or real property given by a lessor to another person (usually called the lessee or tenant) for a fixed or indefinite period of time, whereby the lessee obtains exclusive possession of the property in return for paying the lessor a fixed or determinable consideration (payment). There are two general types of leases, operating leases and capital leases. Operating leases are short-term (five years or less) instruments offered by equipment manufacturers that can frequently be canceled by the lessee before the term expiration. Maintenance and insurance are typically included in the operating lease payments. Capital leases, also called finance leases, generally extend over a longer term equal to the useful economic life of the asset and they cannot be canceled by the lessee. Lease financing is available through banks, finance companies, specialized leasing companies, and equipment manufacturers and retailers. Leasing could be a very efficient way for environmental firms to acquire and use equipment for environmental cleanup or other purposes that may soon become obsolete. Lease payments can be deducted as business operating expenses and can be written off faster than depreciation of owned equipment.

Reference for Further Information: Entrepreneur.com: http://www.entrepreneur.com/money/howtoguide/article52720.html.

Mezzanine Financing

Description: Mezzanine financing is a hybrid of debt and equity financing that is often used to finance the expansion of existing companies. To compensate for the greater risk involved in their usually junior or unsecured status, mezzanine loans typically carry interest rates one to three percentage points above senior, secured loans and include an equity kickers such as stock warrants or convertibility. Mezzanine financing tends to be less restrictive than bank debt and less expensive than equity capital. Companies that have outgrown the start-up phase but are not yet large enough for traditional corporate financing are candidates for mezzanine financing if they have strong management, identifiable growth opportunities, and stable or growing cash flow. Mezzanine financing typically lasts five years and it is sometimes used to finance as much as $100 million. Mid-sized companies providing environmental protection related goods and services may find mezzanine financing to be a good source of capital if their needs are a good match. Some Small Business Investment Companies (SBIC's) specialize in mezzanine financing.

Reference for Further Information: U.S. Small Business Administration Website: http://www.sba.gov/. Contact the Small Business Administration Answer Desk at 1-800-U-ASK-SBA (1-800-827-5722) for the contact information of your nearest Small Business Development Center, which can provide information on the use of mezzanine financing. See "Small Business Investment Companies" in Guidebook Section 10a.
Investopedia: http://www.investopedia.com/terms/m/mezzaninefinancing.asp.

Microcredit

Description: Microcredit is the extension of small loans called microloans to borrowers who are not considered bankable because they lack collateral, steady employment, and verifiable credit history. Recipients of microloans include people living in poverty and cash poor entrepreneurs. Microcredit originated in developing countries, where it has enabled very impoverished people (mostly women) to start successful businesses. Due to the impressive rate of payback on microloans, many members of the traditional banking industry realize that microcredit borrowers are pre-bankable. Thus, microcredit is gaining credibility in the mainstream finance industry. Microloans have traditionally been offered through intermediaries including community development financial institutions. The Small Business Administration has a microloan program. Businesses in the environmental goods and services industry could use microloans for startup or expansion. The International Economic Development Council offers an "Economic Development Finance Programs" course that covers the topic of micro lending.

Reference for Further Information: International Economic Development Council Website: http://www.iedconline.org/index.php?p=Training_Finance_Programs. See "U.S. Small Business Administration Microloan Program" and "Community Development Financial Institutions Fund" in this section of the Guidebook.
Lovgren, Stefan, "Nobel Peace Prize Goes to Micro Loan Pioneers," October 13, 2006, *National Geographic*, at http://news.nationalgeographic.com/news/2006/10/061013-nobel-peace.html.

National Consumer Cooperative Bank

Description: The National Consumer Cooperative Bank (NCB) is a U.S. government chartered corporation organized under the NCB Act in 1978 and privatized in 1981 as a cooperative financial services company. The Act directs the NCB to encourage the development of new and existing cooperatives. The NCB primarily serves cooperatives and their members in the U.S. and its territories by providing them with loans and technical assistance. The cooperatives served by the NCB provide goods and services, housing, education, health care, and other facilities to their members as the ultimate consumers. Cooperatives are owned by their members; and they can help the small businesses that own them to solve environmental, logistical, and marketing problems. The NCB's affiliates and subsidiaries include: 1.) NCB Capital Impact, a 501(c)3 affiliate that provides services including community lending; 2.) NCB, FSB, a federally chartered thrift subsidiary of NCB; 3.) NCB Financial Corporation, a wholly owned stock subsidiary of NCB; 4.) NCB Financial Advisors, a subsidiary providing independent financial advice to nonprofits nationwide; 5.) NCB Community Works, LLC, which provides affordable housing development services; and 6.) NCB Capital Trust I, a Delaware statuary trust.

Reference for Further Information: See "Cooperatives" in Section 5 of this Guidebook. National Consumer Cooperative Bank Website: http://www.ncb.coop/.

National Credit Union Administration: Community Development Revolving Loan Fund

Description: The National Credit Union Administration's Community Development Revolving Loan Fund (CDRLF) was established by Congress to support small credit unions that serve low income communities. The Office of Small Credit Union Initiatives administers the CDRLF, which is designed to establish, strengthen, and improve operations of credit unions and stimulate economic activities in the communities served by credit unions. The CDRLF does this by making loans and Technical Assistance Grants (TAGs) available to qualifying credit unions. The TAGs may be used for purposes such as building credit unions' internal capacity and enhancing their consumer services. The loans may be used for specific purposes that are different from what the grants are generally used for, such as offering new micro business loan programs and educational loan programs, that are potentially useful for environmental protection related purposes. Environmental protection related initiatives that credit unions could use these loans for include providing educational loans to students in environmental studies fields, and awarding loans to businesses providing environmental protection related services and products. CDRLF loans have fixed interest rates of one percent. Credit unions may receive an aggregate amount of $300,000 in loans. The credit union is usually required to provide matching funds.

Reference for Further Information: National Credit Union Administration Website: http://www.ncua.gov/CreditUnionDevelopment/Programs/FinanceGrants.htm.

Accounts Receivable Financing

Description: Accounts receivable financing, also called account receivables factoring, is the selling of a company's accounts receivable, at a substantial discount, to a factor who assumes the risk of the account debtors and receives cash as the debtors settle their accounts. The company receives its money from the factor when it bills or invoices its customers. The discount given to the factor is usually a percentage of the face value of the accounts receivable, usually between 75% and 90%. The factor usually assumes the credit risks related to the receivables. The factor's main concern is with the creditworthiness of the receivables; so factors will finance companies with poor credit that have creditworthy customers. Accounts receivable financing creates no official debt on the company's balance sheet; however it is expensive and significantly reduces profit margins. Accounts receivable financing may be a viable financing approach for firms in the environmental goods and services industry that have marketable products, but experience substantial lag times between placement of orders and receipt of payment.

Reference for Further Information: Most factors can be found in the Yellow Pages of the local telephone book under "Factors" or "Finance." Banks also have lists of local factors.
BusinessFinace.com: http://www.businessfinance.com/account-receivables-factoring.htm.
Investorwords.com: http://www.investorwords.com/54/accounts_receivable_financing.html.
Invoice Financial Website: http://www.invoicefinancial.com/.

Surety Bonds

Description: A surety bond is a contract among three parties: 1.) the principal, who is the primary party performing the contractual obligation, 2.) the obligee, who is the recipient of the obligation, and 3.) the surety, a company ensuring that the principal's obligations will be performed and assuming liability for nonperformance. Through this contract, the surety agrees to uphold, for the benefit of the obligee, the contractual promises made by the principal if the principal fails to uphold its promises to the obligee. In short, surety bonds are given to protect the obligee against loss in case the terms of a contract are not filled. Bid bonds, performance bonds, and payment bonds are three types of surety bonds. Surety bonds are often used to protect the obligee from liabilities related to environmental pollution and contaminated sites. For example, surety bonds are offered through a risk management company called Global Environmental Partners as a means of addressing issues with long-term pollution monitoring and facility closure. In addition, ACSTAR Insurance Company writes virtually every type of surety bond, and has expertise in environmental, pollution, and remediation type contracts. The National Association of Surety Bond Producers can be consulted for more information.

Reference for Further Information: National Association of Surety Bond Producers Website: http://www.nasbp.org/.
Acstar Insurance Company Website: http://www.acstarins.com/q_and_a.htm.
Answers.com: http://www.answers.com/surety%20bonds. Global Environmental Insurance Partners Website: http://www.globalenvironmentalpartners.com/insurance/index.htm.

U.S. Department of Treasury:
Community Development Financial Institutions Fund

Description: The Community Development Financial Institutions (CDFI) Fund, a program of the U.S. Department of Treasury, promotes economic revitalization and community development through investment in and assistance to community development financial institutions (CDFIs). The CDFI Fund was established by the Reigle Community Development and Regulatory Improvement Act of 1994. The Fund's mission is to expand the capacity of financial institutions to provide credit, capital, and financial services to underserved populations and communities in the U.S. It does this by: 1.) directly investing in, supporting, and training CDFIs that provide loans, investments, financial services and technical assistance to underserved populations and communities; 2.) providing tax credits to community development entities, 3.) providing incentives to banks to invest in their communities and in other CDFIs, and 4.) providing financial assistance, technical assistance, and training to Native American CDFIs and Native American entities proposing to become or create Native American CDFIs. Since it was created, the CDFI Fund has awarded $820 million to community development organizations and financial institutions. With support from the CDFI Fund, CDFIs can increase financing for businesses providing environmental protection related goods or services; and they can help finance Brownfields redevelopment.

Reference for Further Information: U.S. Department of the Treasury Community Development Financial Institutions Fund Website: http://www.cdfifund.gov/. See "Community Development Financial Institutions" in Section 9 of this Guidebook (the Brownfields section).

The U.S. Environmental Protection Agency (EPA)'s Environmental Financial Advisory Board (EFAB) is an independent advisory committee established to advise the EPA on environmental financing challenges facing the nation. Chartered in 1989 and operating under the authority of the Federal Advisory Committee Act (FACA), EFAB provides advice and recommendations on environmental finance issues, options, proposals, and trends to the EPA Administrator and EPA's various program offices to assist the Agency in carrying out its environmental mandates. The activities of the Board include analyzing problems, conducting meetings, presenting findings, and other activities necessary for the attainment of its objectives.

The advice EFAB provides the EPA Administrator is about paying for the growing costs of environmental protection and how to increase investment in environmental infrastructure through the leveraging of public and private resources. EFAB provides advice to EPA based on a cross-media, intergovernmental perspective on environmental finance that integrates environmental and economic goals, while emphasizing cost-effective, risk-based approaches and public-private partnerships.

EFAB has adopted three environmental financing goals:

- Lower the costs of environmental protection by removing financial and programmatic barriers that raise costs and by improving the efficiency of investments needed to close the gap between limited resources and increasing mandates;

- Increase public and private investment in environmental facilities and services as a spur to sustainable development, job creation, productivity, and tax revenues; and

- Build state and local financial capacity necessary to carry out environmental mandates so that gains made to date are secured and further environmental progress can be made.

The Board produces policy and technical reports on a wide range of environmental finance related topics of interest to EPA, focusing on environmental finance issues at all levels of government. EFAB is able to address such a wide range of topics because it is made up of members who provide expertise in many different disciplines. The members of EFAB are appointed by the Agency's Deputy Administrator, and they include independent experts from all levels of government, including state and local governments. Its members are also drawn from the banking, finance, and legal communities; business and industry, academia, non-profit environmental organizations, and various national organizations.

The expertise of EFAB's members allows it to provide advice on cross-media financing. This is of critical importance as the nation has invested billions of dollars in environmental facilities and programs over the last thirty years. Environmental legislation reauthorized or enacted by Congress in recent years has placed significant additional resource requirements on all levels of government, increasing their infrastructure and administrative costs. At the same time, limited budgets and changes in federal tax laws have constrained traditional sources of capital. EFAB serves a unique role, assisting the EPA in providing a credible and significant response to the increasing concerns over "how to pay" for federal and state environmental mandates. The Board has made significant contributions to EPA's efforts to address the critical environmental financing challenges of the 21st Century. To view EFAB publications and other related information, visit the website at www.epa.gov/efinpage/efab.htm.

Gregory Mason, Chief Operations Officer
Georgia Environmental Facilities Authority
Atlanta, Georgia

Karen Massey, Deputy Director
Missouri Environmental Improvement and
Energy Resource Authority
Jefferson City, Missouri

Lindene E. Patton
Senior Vice President and Counsel
Zurich North America
Great Falls, Virginia

Cherie Collier Rice
Treasurer and Vice President of Finance
Waste Management, Inc.
Houston, Texas

Helen Sahi, Director
Environmental Services Department
Bank of America
Hartford, Connecticut

Andrew Sawyers, Program Administrator
Maryland Department of the Environment
Baltimore, Maryland

James Smith
Environmental Finance Consultant
Bozeman, Montana

Greg Swartz, Vice President
Piper Jaffray & Co.
Phoenix, Arizona

Steve Thompson, Executive Director
Oklahoma Department of Environmental
Quality
Oklahoma City, Oklahoma

Sonia Toledo, Managing Director
Merrill Lynch
New York, New York

Dr. Jim J. Tozzi, Director
Multinational Business Services, Inc.
Washington, DC

Justin Wilson, Partner
Waller Lansden
Nashville, Tennessee

John C. Wise
Environmental Finance Consultant
Middletown, California

A. Stanley Meiburg, EFAB Designated Federal Official
National Environmental Protection Agency Liaison
Centers for Disease Control and Prevention
National Center for Environmental Health /
Agency for Toxic Substances and Disease Registry
Atlanta, Georgia

The Environmental Finance Center Network (EFCN) is made up of nine Environmental Finance Centers (EFCs) located at universities throughout the United States. It is the only university-based organization in the country that provides innovative solutions to communities to help manage the cost of environmental protection. The EFCs work with the public and private sectors to address the issue of how to finance environmental protection programs and projects. The EFCs provide services and advice directly to individuals and organizations that voluntarily seek out their services.

The Network provides finance training, education, and analytical services designed around the "how to pay" issues of environmental compliance. Each environmental finance center is affiliated with an EPA region. Since the creation of the first center thirteen years ago, the EFCs have expanded into a Network that comprises nine centers strategically located at major universities in eight EPA Regions:

- University of Southern Maine (Region 1)

- Syracuse University (Region 2)

- University of Maryland (Region 3)

- University of Louisville (Region 4)

- University of North Carolina at Chapel Hill (Region 4)

- Cleveland State University (Region 5)

- New Mexico Institute of Mining and Technology (Region 6)

- Dominican University of California (Region 9)

- Boise State University (Region 10)

The EFCs provide state and local governments and the private sector with training and educational, technical, and analytic assistance on environmental finance (see the description of the EFCN and individual descriptions of the nine EFCs in Guidebook Section 5). The services they provide include training programs and electronic financial planning tools. These services are designed around the "how to pay" issues of environmental compliance. The EFC Network has become a significant force in assisting local governments and small businesses in meeting environmental standards. Many of the EFCs work with specific communities to assist with specialized needs, such as the financing of water and wastewater treatment plants. A central goal of the Network is to help create sustainable environmental systems in the public and private sectors.

Environmental protection goals cannot be met without financing, which is essential to implementing state and local programs. The EFC Network's services are based on the premise that communities want to comply with environmental regulations but often do not know how to pay for them. Many communities, particularly small ones, lack in-house financial expertise. Knowledge about how to fund environmental programs is often limited, especially at the local level. The EFCs help fill this knowledge gap- they know that finance is a critical component of sustainable environmental protection. There is a growing demand for the expertise of the EFC directors and staff, who are on the front lines of financing environmental facilities and services.

The finance centers provide state-of-the-art expertise in areas outside of EPA's core competency of developing and implementing environmental programs. The EFC Network has become a significant force in assisting local governments and small businesses in meeting environmental standards. A central goal of the Network is to help create sustainable environmental systems in the public and private sectors. Sustainable systems have the financial, technical, and institutional resources and capability to operate into the foreseeable future in compliance with environmental requirements and in conformance with generally accepted environmental practices. Paying for environmental protection is an important component of sustainability and continues to be primarily a responsibility of local governments and the private sector.

For their part, the financial outreach services of the EFCs help meet environmental needs by identifying ways of cutting costs, lowering and shifting costs, and increasing private sector investment in environmental systems. The work of the EFCs is an ongoing process and the sum total of the Network's benefits make an important contribution to environmental progress in the United States. Further information about the Environmental Finance Center Network can be found on the EPA Environmental Finance Program's Website at http://www.epa.gov/efinpage/efc.htm.

Region 1 Environmental Finance Center
Sam Merrill, Director
University of Southern Maine
Portland, Maine
Website: http://efc.muskie.usm.maine.edu

Region 2 Environmental Finance Center
Mark Lichtenstein, Director
Syracuse University
Syracuse, New York
Website: http://efc.syracusecoe.org/

Region 3 Environmental Finance Center
Joanne Throwe, Associate Director
University of Maryland
College Park, Maryland
Website: http://www.efc.umd.edu/who.html

Region 4 Environmental Finance Center
Lauren Heberle, Director
University of Louisville
Louisville, Kentucky
Website:
http://cepm.louisville.edu/org/SEEFC/seefc.htm

Region 4 Environmental Finance Center
Jeff Hughes, Director
University of North Carolina at Chapel Hill
Chapel Hill, North Carolina
Website: www.efc.unc.edu/index.html

Region 5 Great Lakes Environmental Finance
Center
Kevin O'Brien, Director
Cleveland State University
Cleveland, Ohio
Website: http://www.glefc.org/

Region 6 Environmental Finance Center
Heather Himmelberger, Director
New Mexico Institute of Mining and
Technology
Albuquerque, New Mexico
Website: http://nmefc.nmt.edu/home.php

Region 9 Environmental Finance Center
Sarah Diefendorf, Director
Dominican University of California
San Rafael, California
Website http://www.efc9.org/

Region 10 Environmental Finance Center
Bill Jarocki, Director
Boise State University
Boise, Idaho
Website: http://efc.boisestate.edu/efc/

The Environmental Financing Information Network (EFIN), a component of the Environmental Protection Agency's Environmental Finance Program (EFP), is an Internet-based service offering electronic access to many different forms of information on environmental finance. The purpose of EFIN is to provide information on financing alternatives for state and local environmental programs and small businesses through several channels:

- The EFP Website: The EFP maintains a comprehensive Website of its reports, software tools, case studies, and financial tools as a public resource. Also on the Website, the financial tools pages provide links to sources within the Environmental Protection Agency and outside the Agency.

- Guidebook of Financial Tools: The Guidebook is a reference work examining a wide range of different tools for financing environmental systems.

- Online Database: The EFIN database is a searchable collection of abstracts representing publications and other materials (e.g., articles, case studies, and reports) about environmental financing.

- Infoline: EFIN operates an information phone line that provides callers with referrals, as well as assistance with accessing and searching the EFIN database. The infoline phone number is 202-564-4994, and the hours of operation are 9:00am-5:30pm (Eastern Standard Time), Monday through Friday (excluding federal holidays). EFIN can also be reached via e-mail at efin@epa.gov.

For more information, visit the Environmental Finance Program's Website at: http://www.epa.gov/efinpage/efin.htm.